International Association of Ge

Ivan I. Mueller, Series Ed

International Association of Geodesy Symposia

Ivan I. Mueller, Series Editor

Gravity, Gradiometry, and Gravimetry

Symposium No. 103

Edinburgh, Scotland, August 8–10, 1989

Convened and Edited by
Reiner Rummel
Roger G. Hipkin

Springer-Verlag
New York Berlin Heidelberg
London Paris Tokyo Hong Kong

Reiner Rummel
Afdeling der Geodesie
Technical University Delft
Thijsseweg
2900 JA Delft
The Netherlands

Roger G. Hipkin
Department of Geophysics
Edinburgh University
JCMB, Mayfield Road
Edinburgh EH1935Z
United Kingdom

Series Editor
Ivan I. Mueller
Department of Geodetic Science & Surveying
The Ohio State University
Columbus, OH 43210-1247
USA

For information regarding previous symposia volumes contact:
Secretáire Général
Bureau Central de l'Association Internationale de Géodésie
140, rue de Grenelle
75700 Paris
France

Library of Congress Cataloging-in-Publication Data
Gravity, gradiometry, and gravimetry : symposium no. 103, Edinburgh,
  Scotland, August 8-10, 1989 / convened and edited by Reiner Rummel,
  Roger G. Hipkin.
    p. cm. — (International Association of Geodesy symposia :
  symposium no. 103)
    Includes bibliographical references.
    1. Gravity—Measurement—Congresses. 2. Geodynamics—Congresses.
  I. Rummel, R. (Reiner), 1945-   , II. Hipkin, R. G.
  III. International Association of Geodesy. IV. Title: Gravity,
  gradiometry, and gravimetry. V. Series.
QB330.G73 1990
526' .7—dc20   90-9506

Printed on acid-free paper

Camera-ready copy provided by the editors.

9 8 7 6 5 4 3 2 1

ISBN-13: 978-0-387-97267-1   e-ISBN-13: 978-1-4612-3404-3
DOI: 10.1007/ 978-1-4612-3404-3

Foreword

A General Meeting of the IAG was held in Edinburgh, Scotland, to commemorate its 125th Anniversary. The Edinburgh meeting, which attracted 360 scientific delegates and 80 accompanying persons from 44 countries, was hosted jointly by the Royal Society, the Royal Society of Edinburgh and the University of Edinburgh.

The scientific part of the program, which was held in the Appleton Tower of the University, included the following five symposia:

Symposium 101	Global and Regional Geodynamics
Symposium 102	GPS and Other Radio Tracking Systems
Symposium 103	Gravity, Gradiometry and Gravimetry
Symposium 104	Sea Surface Topography, the Geoid and Vertical Datums
Symposium 105	Earth Rotation and Coordinate Reference Frames

All together there were 90 oral and 160 poster presentations. The program was arranged to prevent any overlapping of oral presentations, and thus enabled delegates to participate in all the sessions.

The 125th Anniversary Ceremony took place on August 7, 1989, in the noble surroundings of the McEwan Hall where, 53 years earlier, Vening-Meinesz gave one of the two Union Lectures at the 6th General Assembly of the IUGG. The Ceremony commenced with welcome speeches by the British hosts. An interlude of traditional Scottish singing and dancing was followed by the Presidential Address given by Professor Ivan Mueller, on 125 years of international cooperation in geodesy. The Ceremony continued with greetings from representatives of sister societies, and was concluded by the presentation of the Levallois Medal to Professor Arne Bjerhammar. The 125th Anniversary was also commemorated by an exhibition entitled *The Shape of the Earth*, which was mounted in the Royal Museum of Scotland. An abbreviated version of the President's speech and the list of all participants are included in the proceedings of Symposium 102.

A social program enabled delegates to experience some of the hospitality and culture of both Edinburgh and Scotland, as well as provided an opportunity to explore the beautiful City of Edinburgh and the surrounding countryside.

A Scottish Ceilidh on the last night concluded a pleasant week, which was not only scientifically stimulating, but also gave delegates and accompanying persons an opportunity to renew *auld acquaintances* and make new ones.

The International Association of Geodesy and the UK Organizing Committee express their appreciation to the local organizers of the General Meeting, especially to Dr. Roger G. Hipkin and Mr. Wm. H. Rutherford, for their tireless efforts in running the meeting to its successful conclusion.

Commencing with these symposia the proceedings of IAG-organized scientific meetings will be published by Springer Verlag Inc., New York from author-produced camera-ready manuscripts. Although these manuscripts are reviewed and edited by IAG, their contents are the sole responsibility of the authors, and they do not reflect official IAG opinion, policy or approval.

V. Ashkenazi	Ivan I. Mueller
A. H. Dodson	President, International
UK Organizing Committee	Association of Geodesy

Preface

Three centuries after the publication of Newton's *Principia*, gravity remains a subject for vigorous research. Symposium 103 reviewed developments in theory, measurement and global gravity models, with applications ranging from searches for the "fifth force" to a determination of sediment structure at the bottom of a deep ocean trench. The presentations were distributed among four themes, with invited reviews, and contributed talks and posters, twenty of which are included as papers in this volume.

The first theme of the Symposium was *Relativity, Fifth Force & Philosophy*. A discussion of the proper definition and choice of units for gravity-related quantities by Biro, was followed by an introduction to the relativistic theory of gravitation by Grafarend. Three speakers, Eckhardt, Jekeli and Harrison, and two posters by Edge & Oldham and Hipkin & Steinberger, together with prolonged discussion, demonstrated continuing activity and controversy in the search for nonrelativistic deviations from Newtonian gravitation. Their field experiments exploited the very high precision of relative gravity meters now available for comparison of gravity observations with predictions based on the inverse square law.

This led naturally to the second theme of *Gravity Measurement*. Liard and a poster by LaCoste & Valliant dealt with new techniques for very accurate calibration of gravity meters. A letter from Faller, read in his absence, looked forward to a complementary accuracy of 10^{-9} g soon to be available with transportable absolute gravity meters. Carter and a poster by Gemael et al. described their absolute gravity measurements in the United States and Brazil, which are already beginning to approach this vision. However, many important applications of gravity data need regionally or globally complete coverage rather than this degree of accuracy. Brozena described a decade of development of airborne gravity measuring systems which are finally beginning to approach the accuracy and resolution needed for regional applications.

Contributions about satellite gradiometry, with the instrumentation described by Bernard and software for gravity field recovery by Sansò, provided a bridge between measurement and the third theme, *Space Techniques & Global Models*. Gradiometry appears as the best hope for acquiring a uniform coverage of accurate, medium resolution, global gravity data. A promising alternative, limited to somewhat longer wavelengths, arises as a spinoff of the relativity experiment Gravity Probe B: Tapley described the potential for gravity field recovery from tracking the probe with GPS. Shum's talk showed that reanalysis of existing data can still yield improvement in global gravity field models, whereas Knudsen's presentation described how new local gravity data can improve a spherical harmonic model locally without global readjustment. Such tailored spherical harmonic models may solve important outstanding problems with local gravimetric geoid calculations.

The final theme, *Regional Gravity, Geodynamics & Tectonics*, was introduced by Bursa's paper on the gravity field of the Martian satellite Phobos. He predicted its field from the measured topography, so linking a global gravity model to tectonic and geodynamic assumptions; the latter were developed further by a discussion of its orbital evolution under tidal friction. On a more local scale, Okuba developed a theoretical model for the change in gravity potential due to the ground displacements of an earthquake and two posters by Becker and by Demirel & Gerstenecker reviewed the use of very precise relative gravity observations to monitor vertical crustal movements. The talk by Gerstenecker described a microprocessor-based instrumentation for field-based reduction and adjustment of such high precision observations. Both oral and poster presentations by Kono & Furuse illustrated the compilation over the last 90 years of regional gravity data for Japan, integrating land offshore measurements into a single machine-readable database. Further applications of Bouguer anomalies for detailed modeling of local geological structures were covered in a poster by Matsumoto & Hotta, dealing with observations made in a deep ocean submersible in the Toyama Trench, and in a poster by Sastry & Moharir describing a new algorithm for gravity anomaly inversion.

It was encouraging to see a range of contributions which spanned the fields of physical geodesy, geophysics and tectonics and brought together both those concerned with techniques and those seeking solutions to specific problems about the Earth.

Reiner Rummel   Roger G. Hipkin

Contents and Program

Regional Gravity, Geodynamics and Tectonics

WHAT IS "GRAVITY" IN FACT?

P. Biró
Technical University of Budapest
H-1521 Budapest

SUMMARY

The term "gravity" has usually three meanings in geodesy as the weight of a body, the intensity of the gravity field and the acceleration of the free fall. Although geodesy does use the only one unit for gravity, that of the acceleration. Technical physics, esp. mechanics need to distinguish the different meanings of gravity also by using the adequate unit in the SI system. In fact the intensity of the gravity field (the force acting on the unit mass) with the unit N/kg is the generally needed quantity instead of the acceleration. Its use will be suggested also in geodesy and proper units in the SI system will be recommended for gravity potential, geopotential number etc.

x

One of the main goals of geodesy is the representation of the earth's gravity field. In this relation the term "gravity" has - at least - three meanings in geodesy and in other geosciences too. One of them is the force acting on a body at the earth's surface (i.e. its weight).

Another meaning of "gravity" is the intensity of the earth's gravity field (i.e. the force acting on the unit mass). The third generally used meaning of it is the acceleration of the free fall. However geodesy does use the only one unit for gravity, that of the acceleration and provides numeric data either in gals (= 10^{-2} ms^{-2}) or in ms^{-2} in SI.

Haszpra (1984, 1986.a,b) calls the attention to the fact that technical physics esp. mechanics needs a clear distinction of the different meanings of "gravity" also by using their adaquate units in SI. Being in agreement with this opinion we shall review the exact content of these three meanings giving suggestions for their proper use in the following.

Gravity force acting on a body in static equilibrium at the earth's surface is defined in a reference system rotating with the earth usually as the resultant $\vec{F}_g$ of the Newtonian (or gravitational) mass attraction $\vec{F}_N$ between the earth's masses and the body and of the centrifugal force $\vec{F}_R$ of the earth's rotation, as

$$\vec{F}_g = \vec{F}_N + \vec{F}_R = (G \int_{earth} \frac{\vec{1}}{1^3}\, dm + \omega^2 \vec{p})\, m \tag{1}$$

being G Newton's gravitational constant (the gravitational force acting between two unit masses from the unit distance), $\vec{1}$ the distance vector between the body and the mass elements dm of the earth body, ω the angular velocity of the earth's rotation, $\vec{p}$ the distance of the body from the spin axis of the earth and m the mass of the body.

In outer space where the attracted body does not rotate with the earth, the second term in the brackets disappears so gravity force and gravitational force can be interchanged. (Definition (1) postulates that the tidal force has been subtracted from the forces acting really on the body in the nature.)

Gravity force has certainly the dimension of force and its magnitude has the unit 1 N (newton) in SI. Consequently the SI unit for the gravitational constant is N m^2/kg^2.

Gravity force as weight is the force either acted by the body on its support in static equilibrium (at rest) or setting the body in motion, without any support or suspension (dynamic effect). The acceleration of the latter can be computed by Newton's second law if all the acting forces are known.

As indicated by (1) gravity force does depend on the mass of the body, therefore it is not convenient for mathematical representation of the earth's gravity field. Therefore a common "reference body" with unit mass will be chosen. Dividing (1) by the mass m one gets the gravity force $\vec{f}_g$ acting on the unit mass i.e. the intensity of the gravity field or gravity intensity as

$$\vec{f}_g = \vec{f}_N + \vec{f}_R = G \int_{earth} \frac{\vec{l}}{l^3}\, dm + \omega^2 \vec{p} = \vec{g} \ . \tag{2}$$

This has been denoted by $\vec{g}$ further on. ($\vec{f}_N$ and $\vec{f}_R$ are the intensities of the mass attraction and of the centrifugal force of the earth's rotation.)

Similarly one gets the dimension of the gravity intensity as a specific force (related to the mass) and the SI unit of its magnitude as N/kg.

The field of the gravity intensity vectors (with given constant earth's mass) can be uniquely described by the $\vec{g} = \vec{g}\,(\vec{r})$ vector-vector function which can be a suitable mathematical representation of the gravity field.

The gravity force acting on (or the weight of) a body with mass m can be easily computed by known gravity intensity as

$$\vec{F}_g = \vec{g} \ . \ m \tag{3}$$

and its magnitude with the correct SI unit

$$F_g(N) = g(N/kg) \ . \ m \ (kg) \tag{4}$$

independently from state of rest or motion.

This is the quantity involved in all static and dynamic problems of mechanics. This will be needed by engineers in dimensioning of structures, solving tasks of hydrodynamics and also by geodesits or astronoms computing the orbits of natural and artificial celestial bodies etc. This corresponds to the intensity of electric, electro-magnetic or magnetic fields generally used by physics for the representation of these fields.

The only one exception in mechanical tasks is the kinematic description of the free fall of a body (with its related tasks) in which the gravity acceleration is involved.

If the body with mass m in the earth's gravity field will not be supported (or suspended) gravity force $\vec{F}_g$ sets it in motion with the acceleration $\vec{a}$. This latter can be computed by Newton's second law (the principal law of dynamics)

$$\vec{F} = m \cdot \vec{a}. \tag{5}$$

In free fall (without any resistance) the only one acting force is the gravity force $\vec{F}_g$ and the acceleration of the free fall (or gravity acceleration) $\vec{a}_g$ can be computed by (5) inserting (3) for $\vec{F}$. Thus:

$$\vec{g} \cdot m = m \cdot \vec{a}_g \tag{6}$$

and dividing by m

$$\vec{g} = \vec{a}_g \tag{7}$$

or for magnitudes with SI units

$$g(N/kg) = a_g \ (ms^{-2}). \tag{8}$$

This indicates that the acting force generates an acceleration being numerically equal in magnitude with the intensity of the force field in SI. It is to be respected that against numeric equality one quantity is of dynamic and the other of kinematic nature. Attention must be called, that this numeric equality occures only in the case of the free fall. If any other force (for example a resistance) be involved the magnitude of

the resultant acceleration will no more be equal to the numeric
magnitude of gravity intensity!

Therefore it can be stated that the quantity representing earth's
gravity field generally needed by representatives of nature sciences and
engineers is the gravity intensity g (N/kg) instead of the
acceleration of the free fall a_g (ms^{-2}).

If the question arises: "What is really the observable by geodetic
methods - gravity intensity or acceleration of the free fall?" the
correct answer will be: "Both - it depends on observation techniques".

By the "free fall method" the magnitude of the acceleration of the
free fall can be determined (after some needed corrections). This can be
converted into the units of gravity intensity by using Newton's second
law by (8) without any theoretical difficulties.

Both by different kinds of gravity meters and by physical pendulums
the magnitude or at least differencies in magnitude of the intensity of
the gravity field can be determined with the appropriate unit of the
specific force (N/kg in SI).

In the present terminology of geodesy and of geophysics there is usually a failure to show a sharp distinction between "gravity acceleration" and "gravity intensity" (and sometimes also "gravity force"). It
is assumed that in geodesy the unit mass is always meant. This will be
supported by the fact that both in geodesy and in geophysics the only
one unit ms^{-2} (or 10^{-2} ms^{-2} = 1 gal) that of acceleration is used.

But it must not be forgotten that the results of geodesy will be used
by many other (nature) sciences whose representatives need (and SI makes
possible) the sharp distinction of intensity (specific force) and
acceleration.

It is e.g. very difficult to explain why in computing the force acting
on a body at the earth surface in static equilibrium (at rest) as the
product of the mass of the body and the gravity acceleration, because in
state of rest there is not any acceleration. But it needs no explanation
if force (N) is computed as the product of the mass of the body (kg) and
the gravity intensity (the specific force) (N/kg) independently from
state of rest or motion.

We find such a conception that the gravity acceleration field could be
viewed as predicting how the gravity force would act if there were a

particle present in the field to be more indirect.

We suggest to use "gravity" generally in the meaning "gravity intensity" with the proper unit. Giving the numeric values of the gravity (intensity) or gravity anomalies in the units 10^{-2}, 10^{-5}, 10^{-8} N/kg (corresponding to gal, mgal and μgal in acceleration) there is no need to change them (except the unit). This could be a compromise between the present practice and the SI.

The mathematical representation of a force field can be more simple by using the scalar function $W = W(\vec{r})$ for this purpose. If this latter be defined by the condition

$$\text{grad } W = \vec{g} \tag{9}$$

it will be termed gravity potential. If we compute the work L_{12} done on the unit mass between points P_1 and P_2 in the gravity field (respecting (9)) we get

$$L_{12} = \int \vec{g} \cdot d\vec{s} = W_2 - W_1 \tag{10}$$

beeing $d\vec{s}$ the element of the way between P_1 and P_2, (10) defines the dimension of the gravity potential as specific work or work done on the unit mass. The appropriate unit in SI is Nm/kg = J/kg, being g(N/kg) and ds(m).

In this sense the potential of the geoid should be given in one of the forms 6.26×10^7 J/kg = 6.26×10^4 kJ/kg = 62.6 MJ/kg. The geopotential number of a bench mark 100 m above the mean sea level should be expressed as 9.81 N/kg x 100 m = 981 J/kg = 98.1 daJ/kg (dekajoule per kilogram). This way the physical content of both measures as specific work done or energy will be completely clear at a first glance also by laymen and can be compared with other energies. (People are familiar e.g. with the energy content of foods.) The unit daJ/kg corresponds to the unit kilogalmeter and the numeric values used at the present are not to be altered. It can be a compromise between the present practice and SI.

As a consequence of (9) the partial derivative of the potential W with respect to any direction x is the appropriate component g_x (N/kg) of the gravity intensity.

The disregard of the difference between specific force and accelerations leads to difficulties and contradictions in the interpretation of
(9). In the geodetic literature one can find the declaration that the
gradient vector of the gravity potential results in the gravity acceleration. By this statement gravity potential will be connected with
the field of gravity acceleration vectors. At the same time it will be
declared that the dimension of the gravity potential is the work done.
By this the fact that only force (and no acceleration) can produce a
work will be forgotten.

As a conclusion the author puts forward for discussion and
consideration a proposal for a resolution of IAG.

R E S O L U T I O N (Draft)

The International Association of Geodesy,

recognizing the need for correct distinction of the terms "gravity
acceleration" and "gravity intensity" (as specific force) also by the
correct use of their units, and

considering the possibilities of SI on one side and the accustomed
usage of the numeric data on the other side,

recommends that

1. the term "gravity" should be used generally in the meaning "gravity
   intensity" with the notation g and to give the numeric values of
   its magnitude either in the SI unit in N/kg or in 10^{-2}, 10^{-5}, 10^{-8}
   N/kg (corresponding to the units gal, mgal and μgal). The same is
   valid also for "normal gravity" γ and for "gravity anomalies" $\Delta g =
   g - \gamma$, being anomalies of gravity intensities with respect to the
   normal gravity intensities. Listed gravity data (or anomalies) are to
   be given in one of the mentioned units. Newton's gravitational
   constant should numerically be given in the unit Nm^2/kg^2. For the
   geocentric gravitational constant GM the unit Nm^2/kg should be used,

2. the term "gravity" in the meaning "acceleration of the free fall" (or shortly "gravity acceleration") with the notation $\vec{a}_g$ should be used only in the kinematic computations (way, velocity, acceleration or time) of a free fall motion (without any resistance). In this and only in this case should be used the unit gal ($= 10^{-2}$ ms^{-2}) or the SI unit ms^{-2} for numeric values,

3. the term "gravity potential" should be used in the meaning "potential of the gravity intensity" with the dimension specific work done (energy) and with the SI unit J/kg. The same for "normal potential" and "disturbing potential" (or "potential anomaly"). (This case the gradient of the gravity potential results in gravity intensity.) Numeric values of the second derivatives of the gravity potential should be given in the unit 10^{-9} Nkg^{-1} m^{-1} (corresponding to 1 Eötvös),

4. geopotential numbers having the dimension specific work done should be given in the unit daJ/kg = 10 J/kg.

REFRENCES

Biró, P. (1989). An improvement of the terms used for mathematic representation of the earth's gravity field (in Hungarian). Geodézia és Kartográfia 41, 1-6.

Biró, P. (1989). Zum Begriff "Schwere" und zu den SI Masseinheiten. Zeitschrift für Vermessungswesen 114, 209-218.

Haszpra, O. (1984). Do not use the SI: exploit it!
g = 9,81 N/kg. Newsletter, Technical University of Budapest, 2, 4, 15-18.

Haszpra, O. (1986 a). Distinguish the intensity of the gravity field and the acceleration of the free fall (in Hungarian). Geodézia és Kartográfia 38, 99-100.

Haszpra, O. (1986 b). Unit for g. Am.J.Phys. 54, 8, 680.

LEAST-SQUARES COLLOCATION ERROR ESTIMATES IN A TEST OF NEWTON'S GRAVITATIONAL LAW

by

Christopher Jekeli and Anestis J. Romaides
Geophysics Laboratory (AFSC), Hanscom AFB, MA 01731

ABSTRACT

Recently, the Air Force Geophysics Laboratory conducted an upward continuation experiment to test the validity of Newton's law of gravitation. One technique that was used to upward-continue ground data was least-squares collocation (LSC). The predicted standard deviation at 600 m above ground in the LSC estimate was significantly smaller than the observed discrepancy between the upward continuation estimate and the measured value. This indicates either a possible violation of Newton's law or a model error (bias). Although the LSC error estimates of gravity interpolation on the ground are shown to be quite reasonable, it is suggested that a terrain misrepresentation can invalidate the error estimates of upward continuation.

INTRODUCTION

An experiment conducted by the Air Force Geophysics Laboratory to test Newton's law of gravitation by comparing measured gravity on a tower with predictions based on a ground network of data turned out to be an extensive exercise in upward continuation. The gravity measurements on the 600 m television transmission tower near Raleigh, North Carolina, were done with relative ease. The extreme stability of the tower allowed us to make measurements with an estimated (absolute) accuracy of 18 μgal (one sigma). The upward continuation, however, originally thought to be straightforward, is much more complex when striving for such accuracy, where some of the usual approximations no longer hold. The first to be invalidated was the "rule of thumb" that upward continuation requires ground data radially out to about ten times the vertical distance (Heiskanen and Moritz, 1967, p.249) which, of course, assumes a desired accuracy of only 1 mgal. We found that our upward continuation estimates did not change significantly only when data out to about 200 km from the tower were included.

One of the techniques used for upward continuation was least-squares collocation (LSC) because it can readily utilize irregularly distributed data at varying elevations and because it yields an estimate of the accuracy of upward continuation. This accuracy depends on the noise of the ground data, as well as on their density (or lack thereof) and extent from the tower. Thus, LSC also can be (and was) used in an error analysis to define the data requirements of such an experiment. The most important aspect of LSC is the adoption of a covariance model. This requires some prudence especially if the error estimates are to be meaningful. This paper describes the methods used to arrive at a useful model and an analysis performed to demonstrate the accuracy (or precision) of the error estimates.

THE COVARIANCE MODEL

The covariance model was designed for the basic gravity anomalies (i.e., not referenced to a higher-degree gravity field). It was deemed easier to include data further from the tower than to deal with inaccuracies of a reference field to some high degree and order. The modeling was done in the so-called frequency (or wavenumber) domain where functions (summable over all wavenumbers) were fitted to empirical determinations of the power spectral density (psd) of gravity anomalies in the North Carolina area. The usual assumptions of isotropy and stationarity in the covariance function were adopted after some rudimentary analyses showed that, indeed, the anomalous field is neither severly anisotropic nor nonstationary. Any deviations from these characteristics (and there are some, of course) would be accurately reflected in and derived from the data themselves in the process of upward continuation. It is the LSC accuracy estimate, perhaps, that suffers most from the assumption of isotropy and stationarity because it is independent of the data values (it depends only on the data locations).

The spectrum of the covariance function was divided into three domains, representing long wavelengths down to about 600 km (degree 65 in a spherical harmonic expansion), intermediate wavelengths from 600 km to 8 km, and short wavelengths from 8 km to about 50 m. At the long wavelengths, the psd comprised the degree variances (from degree 2 through 65) of Rapp's 360-degree spherical harmonic gravity model. For the intermediate wavelengths, a data base of gravity anomalies (point measurements) was used. The data were obtained from Defense Mapping Agency (DMA) and have an average accuracy of 1 mgal. They cover an area about four degrees in latitude and five degrees in longitude centered on the tower and the average spacing between data points is about 4 km. It was appropriate, therefore, to grid these data at regular intervals of 4 km in the north-south and east-west directions. The one-dimensional spectrum of the covariance function was obtained by first computing two-dimensional periodograms of the anomalies and then averaging these with respect to the direction of the wavenumber vectors. Each profile (in both east-west and north-south directions) of the anomaly grid was adjusted so that the end values matched. Fougere (1985) showed that this technique significantly improves the estimation of power-law psd's by the periodogram method (the psd of the Earth's gravity field approximately decays as an inverse power of wavenumber). The empirical psd thus computed (for the intermediate wavelengths) had, on a logarithmic scale, a mean slope of -3.24.

We supplemented DMA's data base within 5 km of the tower (where only a few DMA data were available) with our own gravity survey consisting of 114 points variably spaced (more sparse further from the tower) in approximately concentric rings about the tower (Figure 1). Unfortunately, the data were not sufficiently dense to obtain good estimates of the psd at the short wavelengths. Empirical psd's determined from differently sized grids yielded inconsistent results. We did have a dense set of elevation data, however, and this provided a reasonable estimate. The gravitational effect of the residual topography (with respect to the mean elevation in the area) was computed on a grid with 200 m spacing in both directions. The resulting psd followed the power law established for the intermediate wavelengths (Figure 2). However, continuing this model with the same exponential slope to infinite frequency yields an unbounded variance in the gravity gradient. Therefore, the model psd was cut off in such a manner as to define a reasonable estimate of the gradient variance. This estimate (of the horizontal gradient of gravity anomalies; about $(87\ E)^2$) was obtained from the sample of gravity anomalies in the inner most ring of data (10 m from the tower) by computing finite differences between consecutive data around the ring and averaging their squares.

Finally, a sum of 9 reciprocal-distance covariance models was fitted to the mean shape of the empirical psd. Each reciprocal-distance model has two parameters (a variance and a correlation length) and is analytically summable over all wavenumbers - it is thus easy to implement computationally.

ANALYSIS OF LSC ERRORS

With this model the upward continued gravity field (as defined by our survey and approximately 200 additional points distributed out to 200 km from the tower) differed significantly from the measured field along the tower by amounts that increased to 300 μgal at the top. The standard errors of about 86 μgal were those obtained formally by the least-squares collocation procedure and showed that the observed discrepancy is, indeed, significant. The question remains, however, whether these error estimates are in fact realistic; i.e., whether the discrepancy between observed and computed gravity is real or an artifact of incorrect modeling (bias error).

To test the reliability of the computed LSC errors, the gravity values in each one of the rings of the inner zone survey were removed from the data base and gravity at each of these points was then estimated by LSC operating on the modified data base. The differences between the estimates of gravity in a ring and the corresponding true values give the true errors. A comparison of these true errors with the formal LSC error estimates showed that, in the average, the true error is slightly less than one estimated standard deviation, and no true error is greater than 2.5 estimated standard deviations. This leads to the speculation that the formal LSC upward continuation errors are even conservative in that they slightly overestimate the true error. On the other hand, one then makes the implicit extrapolation that estimated errors in the upward continuation are as plausible as estimated errors of interpolation on the ground. The latter were tested with apparent rigor, but the former could not be tested unless one dismisses the possibility of a non-Newtonian gravity.

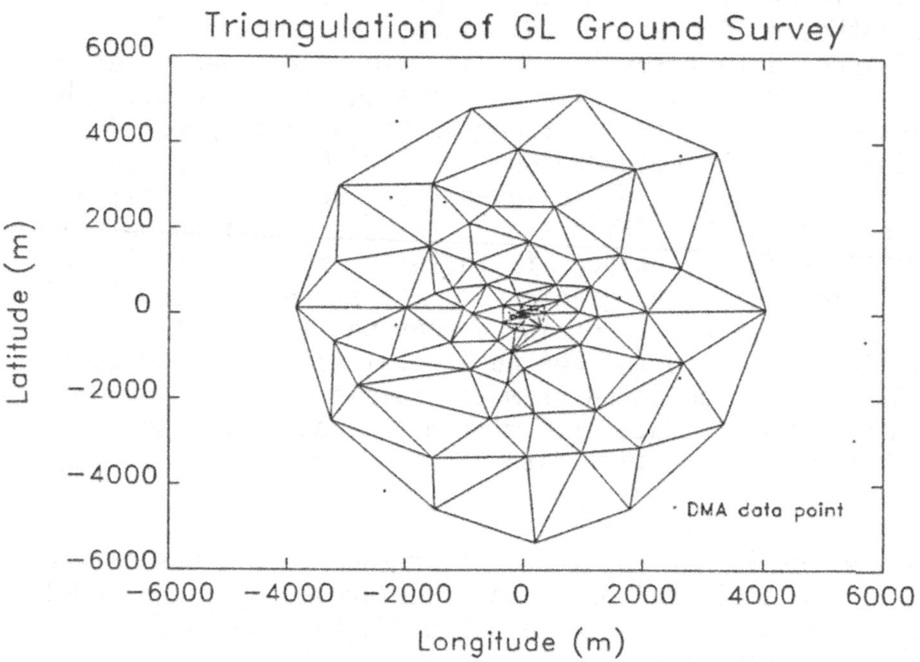

Figure 1: Data points of supplemental survey in vicinity of tower.

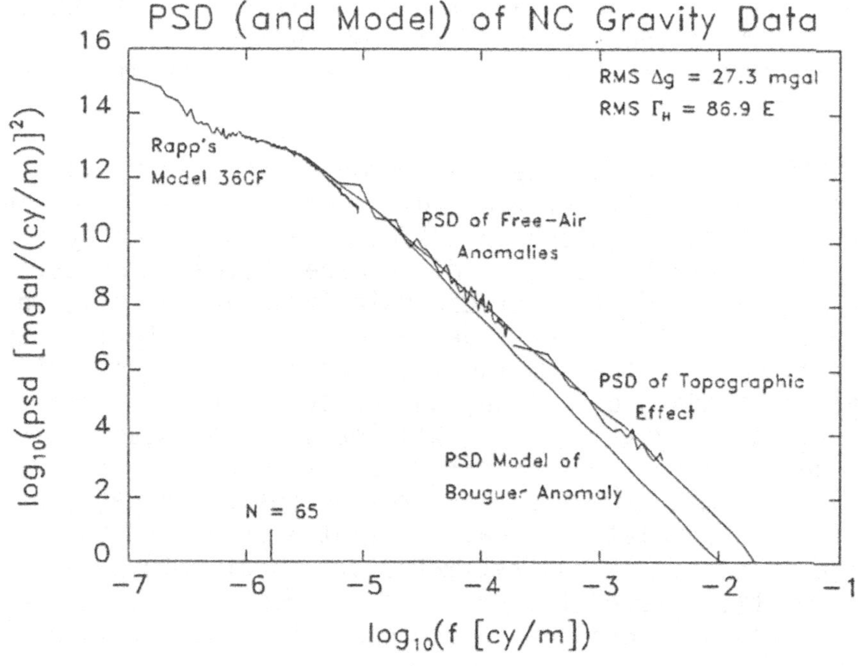

Figure 2: PSD of gravity field in central North Carolina.

NEW RESULT AND ITS CONSEQUENCES

Discrepancies between upward continued estimates and measurements aloft reported earlier by Romaides et al. (1989) were considerably higher (500 μgal) than those stated above. They were based on an inner zone of gravity data containing only 77 points which (as it happened) did not adequately represent the low-elevation areas around the tower. Thus accurate representation of the terrain was found to be a critical problem. Even the augmented data set (114 points) is still biased towards high elevations, where the difference between mean elevations of the data points within a ring and the actual mean elevation for that ring can reach two to three meters (Figure 3).

To settle the question of terrain effect, we divided the upward continuation problem into two stages. First, terrain-corrected anomalies were computed using a detailed digitization of the terrain. These anomalies were treated in a manner similar to the free-air anomalies in an upward continuation by LSC, where now a covariance model for Bouguer anomalies was used instead (this model was derived from corresponding Bouguer anomalies of the DMA data base). In the second step, the effect of the terrain on gravity at heights along the tower was computationally added back. With this two-step procedure most of the originally observed discrepancy between predicted and measured gravity along the tower disappeared.

Does this new result invalidate the analysis of LSC errors described above? Certainly the conclusion of the analysis with respect to the formal upward continuation errors is wrong. Now, since we can not find evidence for non-Newtonian gravity on the basis of this experiment, the true error in upward continuing free-air gravity anomalies (300 μgal) amounts to 3.5 estimated standard deviations (86 μgal). This differs considerably from the ratio of true to estimated errors obtained from gravity interpolation on the ground. Clearly this inconsistency relates to the apparent misrepresentation of the topography by our gravity survey, i.e., the misrepresentation of the very fine structure of the gravity field. The topography in North Carolina is quite flat (mostly under 20 m to 30 m fluctuations); therefore, the topographic effect on gravity is relatively small and is dominated by the Bouguer-slab component. One possible reason for the apparent incongruity between interpolation errors (on the ground) and upward continuation errors is the following. The correlation distance of the vertical component of the topographic effect on the ground, where its total gravitational pull is primarily horizontal, is much smaller than aloft where the vertical component plays a stronger role. Thus interpolation errors are dominated by other errors, since the apparent terrain misrepresentation biases have a relatively small effect. The upward continuation errors (at heights up to 600 m), on the other hand, are more significantly influenced by errors (e.g., biases) in the very short wavelengths. Although LSC incorporates the amplification of correlation distance with altitude, it cannot account for nonstochastic errors, such as terrain misrepresentation biases. In their presence, LSC gives misleading error estimates.

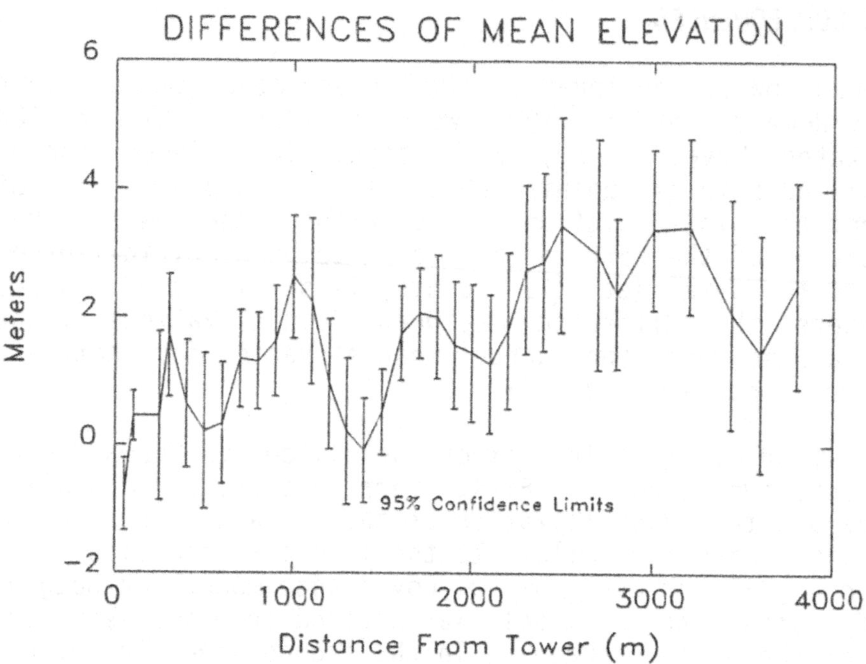

Figure 3: Differences between terrain as modeled by 114 survey points and as represented on a USGS topographic map. The confidence limits refer to the estimation of the azimuthal mean using the survey points.

SUMMARY

In conclusion, LSC has, in general, worked well in processing the data for this experiment, certainly no worse than any other technique. The extreme care applied in designing the covariance model has, at least, not yielded unreasonable error estimates for interpolating gravity on the ground (whether such care was really necessary, is still an open question). The errors of upward continuing free-air gravity anomalies were, however, grossly underestimated. It is thought that the lack of adequately representing the topography by the ground gravity survey is responsible for this result. Conversely, one might argue that it is the covariance model which is in error at the short wavelengths. But the topographic misrepresentation is a systematic (bias type) error, whereas the covariance model describes a presumably stochastic variation in the gravity field. Thus, it is unlikely that a reasonable adjustment to the model at the very high frequencies would yield "correct" upward continuation error estimates in this case. One must first eliminate the systematic error introduced by the terrain misrepresentation.

REFERENCES

Fougere, Paul F. (1985): On the Accuracy of Spectrum Analysis of Red Noise Processes Using Maximum Entropy and Periodogram Methods: Simulation Studies and Application of Geophysical Data. _Journal of Geophysical Research_, vol.90, no.A5, pp.4355-4366.

Heiskanen, W.A. and H. Moritz (1967): _Physical Geodesy_, W.H. Freeman, San Francisco.

Romaides, A.J., C. Jekeli, A.R. Lazarewicz, D.H. Eckhardt, and R.W. Sands (1989): A Detection of Non-Newtonian Gravity. _Journal of Geophysical Research_, vol.94, no.B2, pp.1563-1572.

VALIDATION OF THE INVERSE SQUARE LAW OF GRAVITATION USING THE TOWER AT ERIE, COLORADO, U.S.A.

C. Speake and J. Faller
Joint Institute for Laboratory Astrophysics
University of Colorado, Boulder, Colorado 80309, U.S.A.

and

J. Y. Cruz and J.C. Harrison
Geodynamics Corporation
5520 Ekwill Street, Suite A   Santa Barbara, CA   93111, U. S. A.

INTRODUCTION

Eckhardt et al (1988) point out that gravity measurements on towers provide a straightforward way of testing Newton's inverse square law of gravitation, whose validity in the distance range of 10 - 1000m has been repeatedly questioned during the last decade. Eckhardt et al (loc. cit.) and Romaides et al (1989) have interpreted their measurements on the 600m WTVD tower in North Carolina as evidence for non-Newtonian gravitation. This paper describes a similar experiment using a 300m meteorological tower at Erie on the plains of Colorado some 20km east of Boulder. Gravity was measured at eight levels and, in contrast with the results from North Carolina, these measurements are in excellent agreement with values obtained by upward continuation of surface values using Newton's inverse square law.

THE TOWER MEASUREMENTS

Experiments showed that electrostatically-fedback gravity meters are easier to read, and provide more reproducible results, than meters without feedback under the measuring condition encountered on the tower. The measurements described here were made on 13 round trips up the tower with two electrostatically-fedback LaCoste and Romberg gravity meters. These trips were made on calm nights during the periods of minimum wind speed between midnight and sunrise, and a total of 212 readings were made on the tower distributed between the eight platforms. The meters were calibrated against absolute free-fall determinations covering the range of gravity encountered in the tower and corrections were made for the periodic measuring-screw errors. Platform heights were measured with a laser EDM during each observation and correlation with meteorological data taken on the tower on different nights yielded quite a good determination of its coefficient of thermal expansion $(1.4 \pm 0.1) \times 10^{-5}/°C$. Possible effects of tower sway on its gravity measurements were investigated using analytic techniques, laboratory simulations and, empirically, by searching for a correlation between gravity intervals measured on the different nights and the mean wind speed. It was concluded that sway effects of the magnitude encountered during the measurements are negligible.

In addition, 265 new measurements were made on the surface in the vicinity of the tower - 191 of these within an 800m radius and 70 within 60 meters. In total, there are about 26,000 measurements in a 4° x 5° area, 2640 measurements in a 1° x 1° and 402 measurements within a 10' x 11' area centered on the tower. The gravity interpolations are aided by the fact that both the topography and the Bouguer gravity field are quite smooth within a 20km radius.

THE UPWARD CONTINUATION

The computations of gravity values on the platforms were made in two stages. A global and regional model was constructed using a spherical harmonic expansion of degree and order 8, centrifugal force, and point mass modeling. This model gives an adequate representation of the field more distant than 20km from the tower. Inside this radius the field is considered to be equal to the modeled field plus a residual. The residual field on the topographic surface is estimated within the 20km radius by subtracting the modeled field from the field predicted from the surface measurements. This residual field is continued analytically to a level surface through the base of the tower using Fourier transform techniques (Sideris and Schwarz, 1988) and then continued to the platform elevations by means of the Poisson integral. Gravity was then computed on the tower platforms as the sum of values obtained from the spherical harmonic/point mass model and the upward continued residual field.

Two such computations were made; in the first (model 1), the finest grained mass set was on a 30" x 30" grid, and in the second (model 2) on a 1' x 1' grid. The residual fields in the two cases were, of course, quite different but the platform values predicted as the sums of the modeled and upward continued residual fields are in good agreement. The mass sets are approximately centered on the tower and the masses are at depths equal to their horizontal spacing. The extents and spacings of the mass sets used are given below.

SPACING	EXTENT
5° x 5°	60° x 60°
1° x 1°	26° x 26°
15° x 15°	7.5° x 9°
5' x 5'	4° x 5°
1' x 1' or 30" x 30"	25' x 30'

COMPARISON BETWEEN THE TOWER MEASUREMENTS AND PREDICTIONS FROM THE INVERSE SQUARE LAW

Differences between the tower measurements and predictions from the inverse square law are given in the table below.

Table 1. Comparison between observed gravity values on the tower and Newtonian predictions, in microgals.

Platform Height (m)	Gravity Difference from Base of Tower	Measured - predicted using model 1	model 2	Estimated Standard Errors measurement	prediction
8	-2556	-4	-6	±9	±10
22	-6790	-6	-7	±9	±10
49	-14987	6	7	±10	±11
97	-30003	-15	-9	±12	±12
149	-45912	-3	6	±14	±15
198	-60898	-14	-2	±16	±18
250	-76807	-13	2	±19	±21
295	-90824	4	22	±22	±23

The agreement is excellent and the validity of Newton's inverse square law under the conditions of the experiment is confirmed. The agreement is actually somewhat better than expected from the estimated standard errors. These errors are, however, very hard to estimate and the values given are evidently somewhat conservative.

Acknowledgment.. It is a pleasure to thank the Geodetic Survey Squadron of the Defense Mapping Agency for their active participation in this experiment.

REFERENCES

Eckhardt, D.H., Jekeli, C., Lazarewicz, A.R., Romaides, A.J. and Sands, R.W. (1988). Tower Gravity Experiment: Evidence for Non-Newtonian Gravity, *Physical Review Letters*, **60**, 2567-2570.

Romaides, A.J., Jekeli, C., Lazarewicz, A.R., Eckhardt, D.H. and Sands, R.W. (1989). A Detection of Non-Newtonian Gravity, *Journal of Geophysical Research*, **94**, 1563 - 1572.

Sideris, M. and Schwarz, K.P., (1988). Advances in the numerical solution of the linear Molodensky Problem. *Bulletin Geodesique*, **62**, 59 - 69.

THE INVESTIGATION OF GRAVITY VARIATIONS NEAR A PUMPED-STORAGE RESERVOIR IN NORTH WALES

R J Edge[1] and M Oldham[2]

1 Proudman Oceanographic Laboratory, Bidston Observatory, UK
2 School of Physics, University of Newcastle upon Tyne, UK

Abstract

Continuous gravity measurements have been made for several months near the upper reservoir of the pumped-storage hydroelectricity scheme at Dinorwic, N Wales. A Lacoste and Romberg Earth Tide gravimeter, incorporating electrostatic feedback was installed 80 metres on the landward side of the reservoir and at a height of 10 metres above the maximum water level. Water level changes in excess of 30 metres were observed during the experiment involving the redistribution $\sim 7.10^6$ tonnes of water. A data set in excess of one month has been obtained and analysed in terms of the gravitational effects of the Earth's body and ocean load tides in addition to the gravitational attraction due to the varying water level modelled according to Newtonian gravitational theory. The residual signal was examined in terms of non-Newtonian gravity and possible sources of systematic and random errors considered (e.g. uncertainty in lake geometry, gravimeter calibration and elastic properties of the local rock). The uncertainty in both the permeability and porosity of the rock in the vicinity of the gravimeter is found to be the dominant limiting factor in the overall accuracy of the experiment.

INTRODUCTION

In recent years there has been considerable speculation on the
existence of a "fifth force" of gravity (i.e. does Newton's
gravitational law completely specify the behaviour of gravity over
ranges of zero to infinity?). This fifth force can be expected to
manifest itself in two main ways.

Firstly the dependence of gravitational force on the masses of
attracting bodies might be affected by their chemical composition.
Experiments in this area include those of Fischbach et al (1986), who
effectively started the recent surge in experimental interest by
reanalysing the 1922 results of Eotvos, Stubbs et al (1987),
Adelberger et al (1987), Boynton et al (1987) and Stubbs et al (1989).
All these experiments essentially repeat the work of Eotvos using
modern torsional balances. Within this first approach can also be
included the floating ball experiments of Theiberger (1987), a free
fall experiment of Niebauer et al (1987) and a beam-balance experiment
of Speake and Quinn (1988). The results of these experiments are not
entirely consistent but tend to indicate that the gravitational
attraction between masses is independent of composition.

Secondly there is the group of 'geophysical' experiments which look
for a distance dependence of the gravitational constant $G(r)$ by
comparing observed measurements of gravity with the values of
calculated gravity using Newton's theory and the laboratory value of
the gravitational constant. In the main these experiments have taken
the form of measuring gravity through vertical profiles either
downwards in mines or boreholes or upwards using high towers (see e.g.
Stacey et al 1987, Zumberge et al 1988, Hsui 1987, Romaides et al
1989). These types of experiments again give inconsistent results due
mainly to uncertainties in the density of the material through which
the gravimeter passes, or through problems in the calculation of the
modelled gravity gradient and problems with instrumental effects
(drift, tares etc).

Another class of geophysical experiments is to use continuously
recording gravimeters close to well defined moving large masses such
as water masses in dry docks, hydroelectric storage reservoirs or
tidal basins. This principle has been applied by Moore et al (1988)
and Tuck et al (1988) to a storage reservoir in Australia and, by
Muller et al (1989, in press), to a storage reservoir in the Black
Forest region of Germany. In both cases the constraining geometry has
meant that the measurements are made over ranges of some tens of
metres from the moving mass.

In this paper we report on an experiment aimed at extending the
range of the measurement to ~100 metres near the pumped storage
reservoir at Dinorwic in N. Wales using a well calibrated LaCoste and
Romberg tidal gravimeter. The main difficulty in extending the
distance to the moving water mass is the reduction in the total
observed signal which requires very precise and accurate measurement.
High precision gravimeters have been used previously in tidal gravity
research and have been shown by Edge et al (1986) to be capable of
measuring daily gravity changes ~250 microgals with an accuracy of
better than 0.2%.

The results from a preliminary experiment are compared with model calculations and discussed in terms of the contributory factors (e.g. tides, instrument drift, water density) influencing both the random and systematic error budget of this type of experimental approach.

GRAVITY MEASUREMENTS

During November 1986 LaCoste and Romberg Earth Tide gravimeter ET10 was installed within the Instrument house, approximately 10m above and 80m to the landward side of Marchlyn Mawr reservoir at high water; see Figure 1. The precise position of the gravimeter with respect to a benchmark on the side of the lake was determined to ±3cm horizontally and ±1cm vertically using standard surveying techniques. The gravimeter had previously been calibrated on the short range (3 milligal and 9 milligal) vertical calibration line at Hannover (Kanngieser et al (1983)) and converted to electrostatic feedback from mechanical feedback. Full details of the conversion of ET10, the inclusion of electronic levels and calibration of the measuring screw are given in Baker et al (1981); and Edge et al (1986). A measurement accuracy of 0.2% in amplitude and 0.1° in phase can be achieved using this instrument for gravity signals varying in time over periods of a few hours to a day. Finally the calibrations of the amplitudes and phases of the total gravimeter system (feedback, filters and recorders) were determined, in situ at Dinorwic, by applying step inputs using the calibrated measuring screw both at the start and end of the experiment. Details of the method are given in Baker et al (1981). The difference in calibration between the beginning and end of the experiment was less than 0.1%.

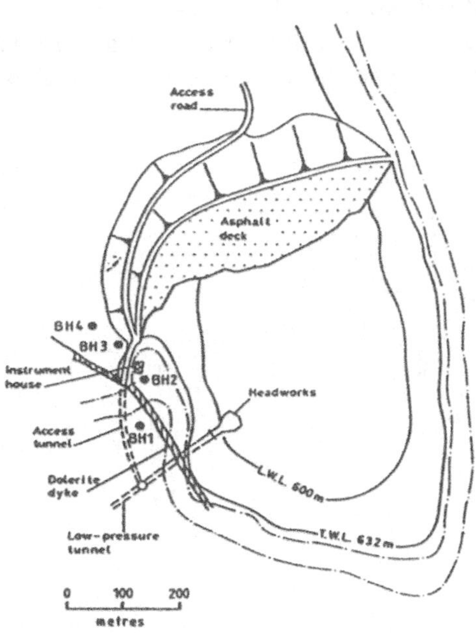

Fig.1 Sketch map of Marchlyn Mawr reservoir showing the location
       of the Instrument House and the boreholes relative to the lake

The gravity data used in this report were digitised, at 30 minute intervals, from the analogue chart obtained from a Phillips PM 8252 two pen chart recorder operating at a speed of 60mmhr^{-1}. Time marks with an accuracy of ±3 secs were superimposed on the record. The incorporation of Cambridge Instruments CS10 range steppers (Baker et al 1981) provides a resolution ~1.0cm microgal^{-1} over a recording range ~600 microgals. The data analysed here are for a 33 day continuous period. Fig. 2 shows the data prior to filtering and shows instrumental drift ~5 microgals day^{-1}.

DATA ANALYSIS

The method adopted in this experiment has been to remain in the time domain and to fit simultaneously the modelled/theoretical effects due to the following
(i) Earth Tides, including ocean tide loading and
(ii) Water mass and mass distribution changes in the reservoir.
The results of this approach are shown in Figures 3a, 3b, 3c and 3d.
To remove instrumental drift the data have been filtered using a standard Xo tidal filter. The resultant time series is shown in Figure 3a.
It has been shown by Baker (1980) that the M_2 tidal loading in the vicinity of Dinorwic is small (~1 microgal) since the site is close to an amphidromic loading node. It is also known, (Baker et al.) that the O_1 loading tide for Europe is small, due to the small amplitude of the O_1 ocean tide in the N.E. Atlantic. Thus to a first approximation the Earth Tide has been modelled using the theoretical gravity body tide (Broucke et al. 1972) with a gravimetric factor of 1.160 for all constituents. This time series is shown in Figure 3b, after filtering using the same standard Xo tidal filter. For different experiments at other locations a much more rigorous approach involving tidal measurements and modelled body tide plus ocean tide loading would be required.

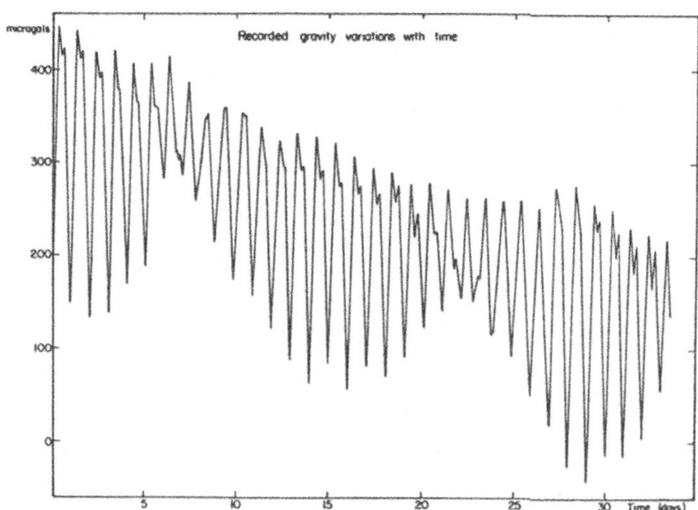

Fig. 2. The observed variations in gravity during the experiment

Modelling of the Lake Signal

The size and shape of the Marchlyn Mawr reservoir has been determined from the final detailed contour map produced after completion of the Civil Engineering work at the site. The 2m spaced contours (estimated accuracy $\pm$ 1m in the horizontal plane) were digitised at spacings appropriate to the structure of the contour and the proximity to the gravimeter. Two different methods were used to compute the Newtonian gravitational attraction of the water.

In the first method the gravitational attraction of a homogeneous body may be written

$$\underset{\sim}{F} = -\nabla\!\int_{V} G\rho/r.dV \tag{1}$$

when G is the Newtonian gravitational constant, ρ is the density of the body and the integral is over the volume of the body. This may be transformed to a surface integral (Spiegel 1982) and may be written as

$$\underset{\sim}{F} = \int_{S} G\rho/r.\underline{n}dS \tag{2}$$

where the integral is now over the surface of the body and $\underline{n}$ is the unit outward normal from the surface element dS.

The surface integration is approximated by using sloping triangular planes formed from the digitised points on the contours. In the case of Marchlyn Mawr the lake signal can be calculated for each 2m contour and intermediate values are obtained by interpolation.

In the second method the contribution from an infinitesimally thin slab of mass at each contour height is calculated and this is integrated in the vertical. This method does not assume plane faces between the contours and thus provides some indication of the errors involved using the first method.

To investigate possible sources of error in these models the contours were re-digitised a number of times with varying degrees of detail. A random Gaussian distribution (mean 0.0m, standard deviation 1.0m) was added to the co-ordinates of all the digitised points to investigate contouring uncertainties.

Comparison among the various models shows a maximum uncertainty in the signal from the reservoir of 0.3%.

RESULTS

Simultaneously fitting the modelled signals due Earth Tides and the varying water mass signal (Figs 3b and 3c) to the observations, (Fig 3a) results in the residual signal shown in Fig 3d (i.e. Observed gravity $-$ (α tides $+\beta$ water signal)) with the coefficients α and β chosen to provide a least squares fit. Performing this least squares fit we obtain $\alpha = 0.999\pm0.001$ and $\beta = 1.072\pm0.004$. The root mean square of the residual time series is 1.3 microgal and is comparable to that obtained from a HYCON standard tidal analysis of Xo filtered data obtained from ET10. The uncertainties quoted for α and β are misleadingly small owing to the cross correlation between the tidal

3a

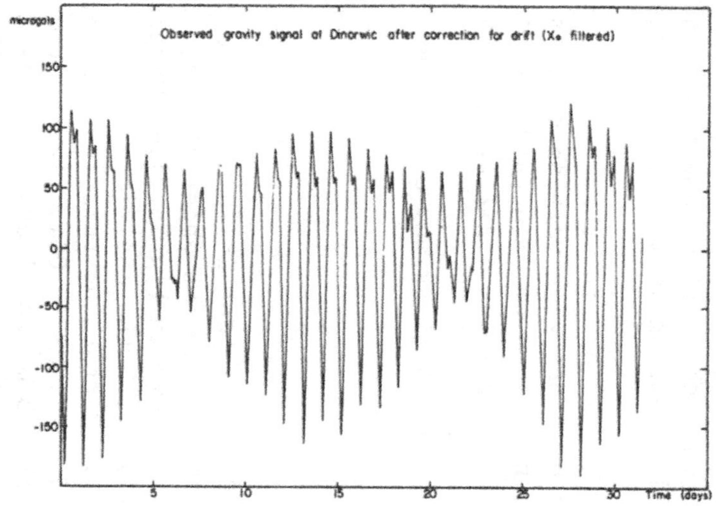
microgals
150
100
50
0
-50
-100
-150
Observed gravity signal at Dinorwic after correction for drift (X₆ filtered)
5
10
15
20
25
30
Time (days)

3b

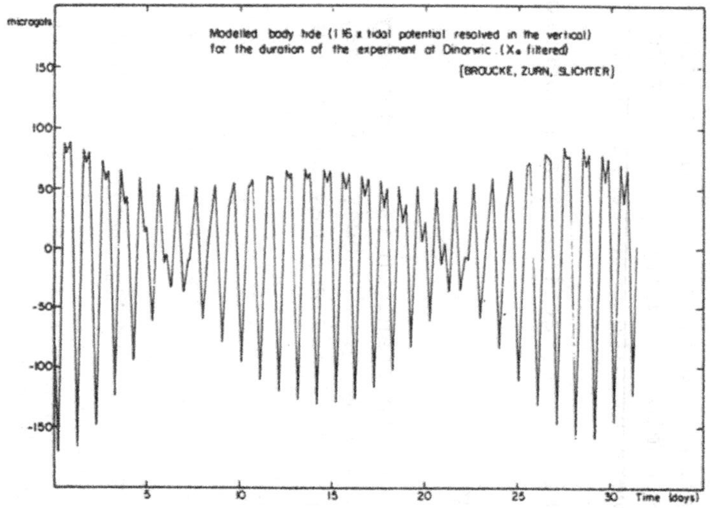
microgals
150
100
50
0
-50
-100
-150
Modelled body tide (1.16 x tidal potential resolved in the vertical)
for the duration of the experiment at Dinorwic (X₆ filtered)
[BROUCKE, ZURN, SLICHTER]
5
10
15
20
25
30
Time (days)

3c

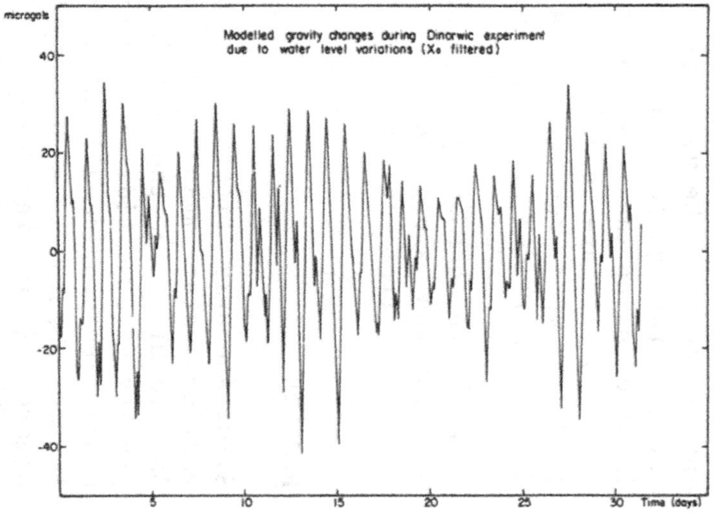
microgals
40
20
0
-20
-40
Modelled gravity changes during Dinorwic experiment
due to water level variations (X₆ filtered)
5
10
15
20
25
30
Time (days)

3d

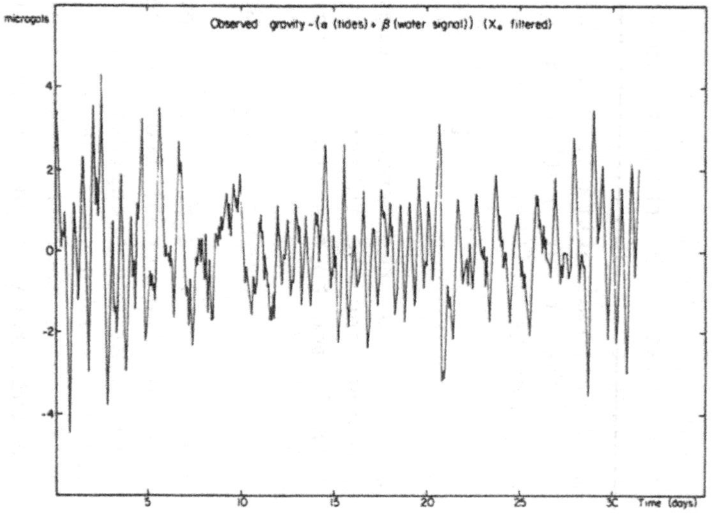
microgals
4
2
0
-2
-4
Observed gravity - (α (tides) + β (water signal)) (X₆ filtered)
5
10
15
20
25
30
Time (days)

signal and the water mass signal. Since the tidal signal is 3 times
larger than the water mass signal the tidal co-efficient of 0.999 is
almost independent of water signal. The water mass co-efficient of
1.072 is slightly more dependent on the tidal signal and if the tidal
signal is changed by 1%, (which for this site is an extremum value)
the water mass co-efficient changes by 3% (i.e. we shall quote here
1.07+0.03). This value is significantly different to a co-efficient
of 1.00 predicted by Newtonian gravity.

SOURCES OF ERROR

The large discrepancy between Newtonian theory and our observed value
either indicates a systematic error or some very interesting physics!
Sources of possible error which have been investigated include

    (i)   Position of gravimeter
    (ii)   Air pressure variations
    (iii)   Water temperature and density variations
    (iv)   Depression of the ground by loading of the lake
    (v)   Uncertainties in lake geometry
    (vi)   Porosity and permeability of the reservoir banks

In this short paper it is not appropriate to expand in detail the
calculations and measurements we have undertaken to determine the
expected sizes of each of these possible error sources. To summarise,
each of the error sources (i) to (v) listed above contribute an error
of less than 0.3%. Together with the uncertainties in the tides and
the modelled geometry of the lake, the large observed discrepancy
(~7%) can not be explained.

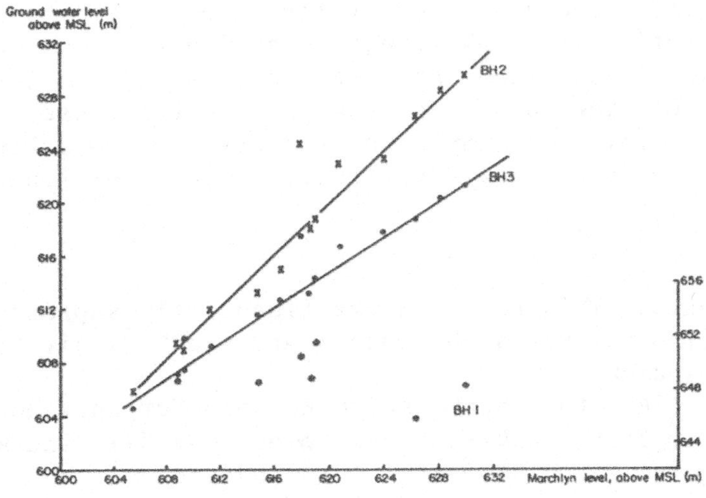

Figure 4   Plots of the variation in water table with reservoir
level in boreholes in the vicinity of the instrument hut

To study the effect of water movements outside the bounds of the modelled water volume, water level observations were made at a number of boreholes in the vicinity of the lake close to the instrument hut. The location of these boreholes is shown in Figure 1 and the observations shown in Figure 4.

From the data of BH3 and BH2 it is apparent that water in the lake is penetrating deeply into the banks and causing large changes in water table directly beneath the gravimeter. The data from BH1 indicate that this penetration does not proceed beyond the igneous dolerite dyke shown in Fig.1.

To quantify the effect of this extra water on the gravimeter it was assumed that the impermeable banks of the reservoir consisted of the dam wall extended along the direction of the access tunnel and the igneous dyke. The extra volume thus enclosed was assumed to be highly permeable and a porosity chosen such that the re-modelled lake signal would reproduce the observations with a Newtonian gravitational co-efficient of 1.00. A value ~1% porosity in this extra volume was sufficient to reduce β to 1.00. A porosity value of 1% is well within acceptable limits for a weathered slate (Clark 1966).

CONCLUSIONS

The gravity measurements at Dinorwic have shown it is possible to achieve good gravity measurements in the vicinity of large moving masses of water.

The interpretation of the signals is highly dependent on the local situation and it has been shown by ourselves and other authors that, for an appropriate geometry, uncertainties in the signals due to Earth Tides, tidal loading, instrumental drift, self loading of the lake, water density and lake contouring may be confined to within acceptable limits for testing the so called "fifth force".

For the particular local situation at Dinorwic the accuracy of the experiment is constrained by the uncertainty in the porosity and permeability of the banks of the lake in the close vicinity of the measurement. This dominant uncertainty in the calculation of the total error prevents a new determination of G over length scales ~100 metres.

Acknowledgements. This research was financially supported by the U.K. Natural Environment Research Council and the U.S. Air Force Office of Scientific Research.

We should like to thank Keith Runcorn for introducing us to the problem and to Trevor Baker, Frank Lowes, Max Hill and Mike Gross for their help and encouragement. In addition we should also like to thank the station manager and the staff of Dinorwic Power station for their use of the facility and in particular we thank Peter Murphy for providing the data on water levels. Roger Hipkin is thanked for providing software for the calculation of the Newtonian signal from the lake.

Graham Jeffries is thanked for his help in the installation and maintenance of the experiment and Neil Hanson is thanked for the analysis of the calibration steps.

Finally we acknowledge the advice and encouragement of Walter Zurn who provided us with full details of his experiences with similar measurements at Hornberg reservoir, West Germany prior to publication.

REFERENCES

Adelberger, E.G., Stubbs, C.W., Rogers, W.F., Raab, F.J., Heckel, B.R., Gundlach, J.H., Swanson, H.E., Watanabe, R. (1987). New constraints on composition-dependent interactions weaker than gravity, Phys. Rev. Lett., **59**, 849-852.

Baker, T.F. (1980). Tidal gravity in Britain: tidal loading and the spatial distribution of the marine tide, Geophys. J.R. Astron. Soc., **62**, 249-267.

Baker, T.F., Edge, R.J. and Jeffries, G. (1981). High precision tidal gravity, Scientific report to U.S. Air Force AFGL-TR-81-0339, 26pp.

Baker, T.F., Edge, R.J. and Jeffries, G. (1989). European tidal gravity. Geophysical Research Letters (in press).

Boynton, P.E., Crosby, D., Ekstrom, P., Szumilo, A. (1987). Search for an intermediate-range composition-dependent force, Phys. Rev. Lett., **59**, 1385-1389.

Broucke, R.A., Zurn, W.E. and Slichter, L.B. (1972). Lunar tidal acceleration on a rigid Earth, Geophysics Monogr. Ser. AGU **16**, 319-324.

Clarke, S.P. (Editor), (1966). Handbook of Physical constants. Memoir **97**, Geological Soc. of America.

Edge, R.J., Baker, T.F. and Jeffries, G. (1986). Improving the accuracy of tidal gravity measurements. In, Proc. 10th Int. Symp. on Earth Tides, Consejo Superior de Investigaciones Cientificas, Madrid, 213-219.

Fischbach, E., Sudarsky, D., Szafer, A., Talmadge, C., Aronson, S.H. (1986). Reanalysis of the Eotvos experiment. Phys. Rev. Lett., **36**, 3-6.

Hsui, A.T. (1987). Borehole measurements of the Newtonian gravitational constant, Science, **237**, 881-883.

Kanngieser, E., Kummer, K., Torge, W., Wenzel, H.-G. (1983). Das Gravimeter-Eichsystem Hannover, Wiss. Arb. d. Fachrichtg. Vermessungswesen d. Univ. Hannover, **120**, 95p.

Moore, G.I., Stacey, F.D., Tuck, G.J., Goodwin, B.D., Linthorne, N.P., Barton, M.A., Reid, D.M., Agnew, G.D. (1988). A determination of the gravitational constant at an effective mass separation of 22m, Phys. Rev. D, **38**, 1023-1029.

Niebauer, T.M., McHugh, M.P., Faller, J.E. (1987). Galilean test for the fifth force, Phys. Rev. Lett., **59**, 609-612.

Romaides, A.J., Jekeli, C., Lazarewicz, A.R., Eckhardt, D.H., Sands, R.W., (1989). A detection of non-Newtonian gravity, J. Geophys. Res., **94**, 1563-1572.

Speake, C.C., Quinn, T.J. (1988). Search for a short-range, isopin-coupling component of the fifth force with use of a beam balance, Phys. Rev. Lett., **61**, 1340-1343.

Stacey, F.D., Tuck, G.J., Moore, G.I., Holding, S.C., Goodwin, B.D., Zhou, R. (1987). Geophysics and the law of gravity, Rev. Mod. Phys., **59**, 157-174.

Stubbs, C.W., Adelberger, E.G., Raab, F.J., Gundlach, J.H., Heckel, B.R., McMurry, K.D., Swanson, H.E., Watanabe, R. (1987). Search for an intermediate-range interaction, Phys. Rev. Lett., **58**, 1070-1073.

Stubbs, C.W., Adelberger, E.G., Heckel, B.R., Rogers, W.F., Swanson, H.E., Watanabe, R., Gundlach, J.H., Raab, F.J. (1989). Limits on composition-dependent interactions using a laboratory source: is there a "fifth force" coupled to isopin?, Phys. Rev. Lett., **62**, 609-612.

Thieberger, P. (1987). Search for a substance-dependent force with a new differential accelerometer, Phys. Rev. Lett., **58**, 1066-1969.

Tuck, G.J., Barton, M.A., Agnew, G.D., Moore, G.I., Stacey, F.D. (1988). A lake experiment for measurement of the gravitational constant on a scale of tens of meters, (preprint).

Zumberge, M.A., Ander, M.E., Lautzenhiser, T., Aiken, C.L.V., Parker, R.L., Ferguson, J.F., Gorman, M.R. (1988). Results from the 1987 Greenland G experiment, EOS Trans. Amer. Geoph. Union, **69**, 1046.

TESTING NEWTON'S LAW IN THE MEGGET WATER RESERVOIR

Roger G Hipkin & Bernhard Steinberger
Department of Geology & Geophysics
University of Edinburgh, Edinburgh EH9 3JZ, United Kingdom

SUMMARY

Gravity was measured at seven levels inside a submerged tower in Megget Water Reservoir, Scotland, in an experiment to test the inverse square law of gravitation over length scales from decimetres to tens of kilometres. The effects of Earth rotation, the global mass distribution, natural topography and man-made structures must be predicted at points outside the solid Earth, in order that that part of the measured attraction which is due to water in the reservoir can be separately identified. This attraction can be related to the Newtonian gravitational constant because, unlike natural rock masses, the density of water can be reliably predicted. This paper describes techniques aiming to determine all necessary corrections with a differential accuracy of 5 $\mathrm{nm.s}^{-2}$ over a 50 m vertical interval in order to exploit the 10 $\mathrm{nm.s}^{-2}$ accuracy available from LaCoste & Romberg gravity meters.

INTRODUCTION

'Big G' experiments using water as the attracting test mass fall into two classes: those where the measurement point is fixed and a mass of water moves (for example the experiment in which a dry dock is filled, or the varying water level in a pump storage reservoir - see Edge & Oldham in this volume), and those, like the Megget Water experiment, where the attraction of a fixed mass of water is observed at different distances. The latter are inherently less tractable because the effect of all other masses, most particularly the topography, also changes as the observing position is moved and the contribution of the water must be isolated from them. The only redeeming feature of the second class of experiment is that the attraction of the water can amount to several milligals rather than only a few microgals, so that, in principle. greater accuracy in estimating the gravitational constant can be achieved. Practical realisation of this potential requires accurate and reliable computation of the attraction of all other masses.

CORRECTIONS FOR THE GLOBAL GRAVITY FIELD

Earth rotation and normal gravity

By first removing a reference gravity field, the main analysis need only deal with the anomalous gravity field, which is free of the effects of Earth rotation for measurements made at rest in a frame rotating with the Earth. Heiskanen & Moritz (1967) (H&M) give a closed formula on p 76 for 'normal gravity', γ, determined on the ellipsoid. This experiment evaluates it using the set of adopted parameters which define the Geodetic Reference System 1980. Comparison with recent observed values implies that the scale of GRS80 normal gravity is uncertain by less than 4 parts in 10^7. (Because this investigation is concerned with gravity differences, the accuracy of the GRS80 gravity datum, only about 3900 nm.s^{-2}, is not relevent to this experiment.)

Formulae used to determine the variation of normal gravity with height along the ellipsoidal normal need some caution because most are given and derived as series expansions and the accuracy required here is unusually large. H&M (p70) give a closed formula due to Bruns for the first derivative of normal gravity along the ellipsoidal normal. After some manipulation, it is

$$\frac{\partial \gamma}{\partial h} = \frac{-\gamma}{a(1-e^2)^{\frac{1}{2}}} (1-e^2\sin^2\phi)^{\frac{1}{2}}(2-e^2-e^4\sin^2\phi) - 2\omega^2 \tag{1}$$

Higher order derivatives generally involve approximations but a closed formula for them is derived here, using confocal ellipsoidal polar coordinates (u,β,λ) (H&M, sections 1.19 & 1.20). A curve following the ellipsoidal normal is obtained by varying u, essentially the semi-minor axis of the ellipsoids, and holding the other two coordinates fixed. Consequently, closed formulae for higher derivatives of normal gravity can be found by recasting Brun's formula (eq 1) in ellipsoidal polar coordinates and partially differentiating with respect to the minor axis. For example,

$$\frac{\partial^2\gamma}{\partial h^2} = \frac{1}{\gamma}\left(\frac{\partial\gamma}{\partial h} + 2\omega^2\right)\frac{\partial\gamma}{\partial h} - \gamma\frac{\partial}{\partial h}\left[\frac{1}{\gamma}\left(\frac{\partial\gamma}{\partial h} + 2\omega^2\right)\right] \tag{2}$$

The first term is obtained from Brun's formula and the second is given after manipulation to and from ellipsoidal coordinates by the *exact* expression

$$\gamma\frac{\partial}{\partial h}\left[\frac{1}{\gamma}\left(\frac{\partial\gamma}{\partial h} + 2\omega^2\right)\right] = \frac{\gamma}{a^2(1-e^2)^2}\left((2 - e^2 + e^4) - (7 - e^2 + e^4)\sin^2\phi\right.$$
$$\left. + 8e^4\sin^4\phi + (1 - 2e^2 + 2e^4)\sin^6\phi\right) \tag{3}$$

Thus normal gravity can be calculated from a Taylor expansion about its value on the ellipsoid. For the location of the tower ($\phi = 55°.49508367$),

$$\gamma(h) = 9815494767. - 3084.80777h + 0.000722939h^2 - 0.0000000303h^3 \ldots \text{nm.s}^{-2} \tag{4}$$

The highest observation point in the tower lies 336.00m above sea level, corresponding to an ellipsoidal height of about 390m. The quadratic term contributes 26.6 nm.s^{-2} to the gravity difference between the top and bottom of the tower but the cubic term is negligible at 0.06 nm.s^{-2}. However, this computation of the gravity difference is a very significant 87 nm.s^{-2} smaller than the one given by the standard factor of 3086 nm.s^{-2}/m adopted by the Bureau Gravimetric International.

The external effect of the anomalous gravity field

If the validity of the inverse square law is assumed, Laplace's equation holds. Predicting the external gravity field then only involves a surface integral of gravity anomalies: no assumptions are needed about the internal distribution of density.

However, integration is not trivial. Measurements of the anomalous gravity field are only available at discrete points on the Earth's topographic surface, and the integral can only be evaluated reliably if the anomalous field varies smoothly enough for valid interpolation between observations. The anomalous gravity field varies both because of lateral variations in subsurface rock density and because of the complex, non-ellipsoidal shape of the Earth's surface. Although the long wavelength components of the anomalous gravity field can be satisfactorily modelled as a finite spherical harmonic series, the effect of local topography generally predominates at short wavelengths. If great accuracy is required, this frustrates any attempt at direct integration for all but topographically benign regions: gravity observations cannot practically be made close enough together to recover short wavelength variations in ground height.

The difficulty can usually be circumvented by removing the calculated gravitational attraction of a model which represents the topographic geometry but assigns it a constant density. The difference between the observed 'free air' gravity anomalies and the attraction of the model topography are then related only to deviations from the assigned topographic density, together with lateral density variations at deeper levels. The short wavelength component of these 'complete Bouguer anomalies' is very greatly reduced compared with 'free air anomalies'. Consequently, the distance over which the anomalous gravity field can be adequately interpolated is increased and the maximum distance separating gravity observations become feasible.

The external effect of the anomalous gravity field was thus computed in three parts: first, the effect of long wavelength, global components; secondly, the effect of a local topographic model, and, thirdly, the contribution from differences between the global model and locally measured gravity anomalies, corrected for the attraction of local topography.

Long wavelength gravity anomalies

The OSU86E spherical harmonic model of the anomalous gravity field (Rapp & Cruz, 1986) was used to calculate the global component of the anomalous gravity field. The model is complete to degree and order 360 and so, in principle, represents wavelengths longer than about 110 km.

The model field was calculated at 50m intervals along the ellipsoidal normal for heights between 0 and 700m above the ellipsoid. (The ellipsoidal height of the centre of the tower is about 364m.) This procedure required a high computational precision because each point requires the summation of more than 130000 terms. The anomalous gravity field was found to be

$$\Delta g_{OSU86E} = 31945.54 - 0.35926\,h + 0.00000342\,h^2 \ \mathrm{nm.s^{-2}}$$

The effect of the long wavelength components of the anomalous gravity field was thus a decrease of 17.9 $\mathrm{nm.s^{-2}}$ between the lowest and highest station in the tower.

GEOMETRICAL MODELS OF LOCAL STRUCTURES

Overview

Megget Water reservoir lies in a valley with the characteristic U-shaped cross-section of a glacial feature, having been deeply cut into Silurian greywackes during the Pleistocene. Nearby hills on its flanks rise 500 m above the valley floor and, although the draw-off tower in which the gravity measurements were made lies near the centre of the valley, analysis of the gravitational effect of topography requires special care. For all computations, a provisional value of 6.673 10^{-11} $\mathrm{m^3.s^{-2}.kg^{-1}}$ was adopted for the gravitational constant.

Distant topography

The topographic model for distant topography was constructed from 1km square vertical prisms, whose height was estimated manually from 25' (7.6m) contour maps. These mean elevations were refered to a quadratic surface locally representing the curvature of the geoid. The gravitational attraction was computed from the full expression for a cuboid in an inner zone, with successive zones using the cylindrical sector and then the vertical line element approximations. A near zone, 8 km by 10 km, was excluded from the prism model which was extended to a distance of about 120 km before the incremental contribution differed by less than 5 $\mathrm{nm.s^{-2}}$ between the top and bottom of the tower.

The contour integration program

The gravitational attraction of the remaining parts of the natural topography, together with models for the reservoir embankment, draw-off tower and water were determined using a development of the contour integration algorithm of Talwani & Ewing (1960). This involves analytic integration for the attraction of a horizontal polygonal lamina, combined with numerical integration over the vertical coordinate to find the bulk attraction. Extensive numerical comparisons were made with spheres, cylinders, cuboids, and parabolic domes (for whose gravitational attraction analytic formula exist), in order to find an adequate numerical integration routine and determine the necessary density of contour information.

Oldham (personal communication, 1988) reports that the results of our version of the contour integration program were insignificantly different from those obtained with an anlytical surface integration over the triangular facets of a polyhedron defined by the same digitised contours.

Intermediate topography

Within a region defined by the National Grid coordinates [315 km < Easting < 325 km, 617 km < Northing < 625 km], but excluding an innermost rectangle 750 m by 1025 m, a topographic model was constructed by digitising at 25 m intervals along every 100' (30.5 m) contour on Ordnance Survey maps. In the region later occupied by the western (most distant) end of the reservoir, 25' (7.6 m) contours were digitised.

Local topography

Before construction of the reservoir embankment was begun, control pillars were established on the adjacent hillsides and levelled in to Ordnance Survey bench marks. A photogrammetric survey was then carried out, from which 2 m contours were mapped at a scale of 1:1250. This map, and all subsequent construction work, were referred to a local coordinate system parallel to the National Grid. The precise National Grid coordinates of the origin of the local system were no longer available but comparison of numerous landmarks common to both the local and Ordnance Survey maps determined them with a standard deviation of 0.7 m.

After the dam was complete but before the reservoir was full, a second photogrammetric survey was made and a second contour map produced. There were additional levelled control points on the dam, new access roads and water channels. From these two surveys, a model of the local topography was constructed by digitising at 2.5 m intervals along every 2 m contour on both maps. The natural topography was defined by the first photogrammetric survey, or the foundation of any subsequent earthworks or construction.

The reservoir embankment

The reservoir embankment has a triangular cross-section which, at the centre of the valley, is about 60 m high and about 160 m wide. The embankment was constructed from compacted soil and gravel with a central vertical membrane of asphalt 700 mm thick. At the base of the membrane is a concrete inspection gallery running the length of the dam and set into grouted bedrock. From the control room on the down-stream face of the embankment, a second inspection and access gallery runs beneath the dam to the draw-off tower rising from the reservoir floor. The drinking water aquaduct and overflow spillway lie beneath the floor of this gallery.

The model for what is loosely called the dam includes all made ground: it includes superficial earthworks for embankments for new roads, water channels and landscaping, as well as the reservoir embankment itself. The model is essentially defined from engineering drawings, with additional control from the two photogrammetric surveys.

The water draw-off tower

The concrete tower consists of two concentric circular cylindrical shells mounted on an octagonal base. The outside diameter is 24 m and the walls are 700 mm thick. Drinking water is drawn off on the up-stream side by ten 1400 mm diameter pipes, two at each of five levels at 8 m intervals from the bottom of the reservoir. Where the water enters, the two cylindrical shells are connected by a solid concrete sector but, elsewhere, the cavity between them serves as an overflow spillway. There is a further inlet to the spillway at the base of the tower to provide a controlled volume of water for the river downstream. The surface level and volume flow of drinking water and compensation water are monitored continuously.

Gravity measurements were taken about 1.4 m west of the vertical axis of symmetry of the cylinders on concrete floors at seven levels in the tower. The axis of maximum symmetry was inaccessible. Although the tower geometry in superficially simple, an embedded lift shaft, pipes, stairs and galleries added to its complexity. Ultimately, more than 2 Mbyte of information was used to construct a model for the tower: digital horizontal cross-sections, determined with 1 mm precision from engineering drawings and randomly checked at the 10 mm level with confirmatory tape measurements, were prepared at 200 mm vertical intervals. To recover the detailed shapes of local horizontal structures, additional cross-sections were added at 20 mm intervals where necessary.

Water

The model for water in the reservoir consisted of three parts. The most distant part was defined by 25' (7.6 m) contours from Ordnance Survey maps. The main contribution came from the area of the photogrammetric survey surrounding the tower. Finally, a small but not insignificant effect came from water in the pipework of the tower.

During the gravity observation programme, the water level was 1.72 m below the overflow datum at 334.00 m above sea level. For the most distant part of the model, defined by 25' contours, the effect of water filling the reservoir to 1025', 1050', 1075' and 1100' (335.28 m) was computed and the required effect of water at 332.28 m was found by linear, quadratic and cubic interpolation polynomials. The interpolation errors appeared to lie within 5 nm.s^{-2} at each site. The same interpolation procedure was used for the main contribution defined by the 2 m contour model, again with insignificant errors.

Water entering each of the drinking water inlet pipes is monitored continuously for temperature (to 0.1K), pH and gas content. It was apparent that the water was well mixed, with a top-to-bottom variation of temperature of less than 0.2K about a mean of 7.2°C. The atmospheric pressure was measured with precision surveying barometers with an accuracy of about 0.1bar. The density of water in the reservoir was thus determinable as (0.99988 ± 0.00001) Mg.m^{-3}.

ANALYTIC CONTINUATION OF THE RESIDUAL BOUGUER ANOMALIES

In order to compute the variation with height of the residual Bouguer anomaly, all point gravity observations within the 126 km square [258 km < Easting < 384 km, 560 km < Northing < 686 km] were extracted from an integrated gravity database (Hipkin & Hussain, 1982). Within this area, there are 8796 stations, mostly observed by the British Geological Survey or Edinburgh University. The mean station density of 0.55 km^{-2} was moderately uniform despite the difficult terrain because of helicopter and pedestrian surveys. All data were uniformly reduced, with an adopted terrain density, constant throughout, of 2.7 Mg.m^{-3}. The reduction also adopted a value of 6.673 10^{-11} kg.m^3.s^{-2} for the gravitational constant. Both figures are identical to the ones used to calculate the attraction of the topographic model, so that the combination is independent of these choices. Terrain corrections were completed to at least 22 km and usually to 64 km. Although duplicate and very closely spaced stations had been eliminated from the database, consistency between the two main surveys had previously been demonstrated at 100 - 400 nm.s^{-2} level, compatable with their target accuracy of 500 nm.s^{-2} (0.05 mgal).

The variation of the Bouguer anomaly with height in the tower was found as a by-product of a transformation algorithm described in Hipkin (1988). This Fast Fourier Transform routine was designed to transform point observations irregularly distributed over a topographically irregular surface to a regularly gridded representation on a plane. It was applied to the 126 km square of data with four options: first, the interpolation grid size was varied between 1 and 2 km; secondly, variable marginal tapering and optional linear detrending were applied before Fourier transformation; thirdly, the altitude of the horizontal plane was varied between 0 amd 700 m above sea level, and, finally, the vertical gradient was determined either as a first derivative on the plane or by linear interpolation between values on two horizontal planes. These options generated slightly different estimates for the vertical gradient of the residual Bouguer anomaly: the mean and standard deviation of six estimating procedures were:

$$\Delta g(h) - \Delta g(0) = -(1.88 \pm 0.32)\ h\ \ nm.s^{-2},$$

with a gradient range between -2.39 and -1.51 nm.s^{-2}/m. The option likely to give the best result involved interpolation between planes separated by a 65 m vertical interval covering the tower measurements and used a detrended 2 km grid with a 10% marginal taper; it gave -1.85 nm.s^{-2}/m.

This scatter reflects the fact that the data spacing was not quite close enough in the region around Megget Water to achieve the desired accuracy: one standard deviation in the Bouguer anomaly gradient generates 16 nm.s^{-2} between the top and bottom of the tower.

The effective topographic density

The constant density used in the topographic model was varied until its attraction calculated in the tower most closely approximated that of the combination of the Bouguer anomaly gradient of -(1.88 $\pm$ 0.32) nm.s^{-2}/m and the model using a density of 2.7 Mg.m^3. This defined a locally effective mean topographic density of (2.7082 $\pm$ 0.0014) Mg.m^{-3}.

This figure can be compared with a direct gravimetric density determination in the area. All 196 point gravity observations in a 20 km square surrounding the tower were assumed to generate a Bouguer anomaly described by a second degree polynomial surface. The terrain density was then adjusted in a least squares miminisation of gravity residuals. This gave a local regression density of (2.7094 $\pm$ 0.0030) Mg.m^{-3}. (Note that both of these methods really estimate the product of density and the gravitational constant. The given value of density refers to the adopted value of the gravitational constant quoted above.)

The two methods of estimation are conceptually different: the second method assumes that the topographic density is constant, whereas the first uses data which have not involved this assumption but merely approximate the result as a single effective density. It is however clear that the density really is unusually uniform: substituting the effective density into the topographic model calculation gives residuals against the original values with a standard deviation of only 7.5 nm.s^{-2} at the seven sites in the tower.

It was anticipated that the bedrock had uniform physical properties over a wide area. The Silurian greywacke formation, which some seismic models imply extends down to about 10 - 15 km, also extends over much of southern Scotland and Northern Ireland. Even on this scale, the density varies very little: Hipkin & Hussain (1982) used the regression density technique on 34 separate 10km squares and found an overall average of (2.728 ± 0.045) Mg.m^{-3} from 1688 gravity observations.

In the Megget Water valley, the conditions are particularly favourable: because of glaciation, the rockhead has been exposed from a depth of burial of at least 300 m in geologically recent times and so is virtually unweathered. Nevertheless, finding both estimates of the effective topographic density so similar and so well determined is particularly fortuitous and is crucial to the ultimate success of the experiment. It is almost always possible to find a 'pathological' density distribution which generates a measureable gravity field at some points and none at others. For this reason, any experiment based on predicting the external field from a finite set of point measurements needs recourse to a hypothesis that the field varies 'reasonably' between the observation points. Observable uniformity of the topographic density is prima facie evidence that it does.

GRAVITY MEASUREMENTS

Gravity was measured at 7 sites within the tower and at 14 sites on the reservoir embankment. Analysis of the latter is incomplete and this paper gives a preliminary interpretation based on the tower measurements alone. The observation were made in a symmetrical triple looping sequence with the LaCoste & Romberg gravity meter G-275. This meter has been calibrated against the IGSN71 scale with a precision of 2 parts in 100000 (Hipkin et al 1988). The possibilty of small gear train errors is currently being investigated by comparison with the gravity meters D-145 and D154, which have twin calibrated dials and electrostatic feedback. Given the gravity range of about 9 dial turns (90512 nm.s^{-2}) in the tower, the periodic screw errors are likely to be less than 50 nm.s^{-2}. Network adjustment using robust statistics with *a posteriori* weighting shows that the data have moderately good internal consistency, with a standard error of between 16 and 29 nm.s^{-2} at the tower sites. It is anticipated that additional observation sequences will halve this error.

RESULTS AND DISCUSSION

Figure 1 shows residuals between observed and calculated gravity at each of the 7 tower sites, together with the effect of increasing or decreasing the gravitational constant from the adopted value of 6.673 10^{-11} m^3.s^{-2}.kg^{-1}. Although adjustment would imply an insignificant correction to the laboratory value, the residuals are not only unacceptably larger than the observational errors, but also show systematic deviations in mid-elevations. The rate of variation with height in the upper half of

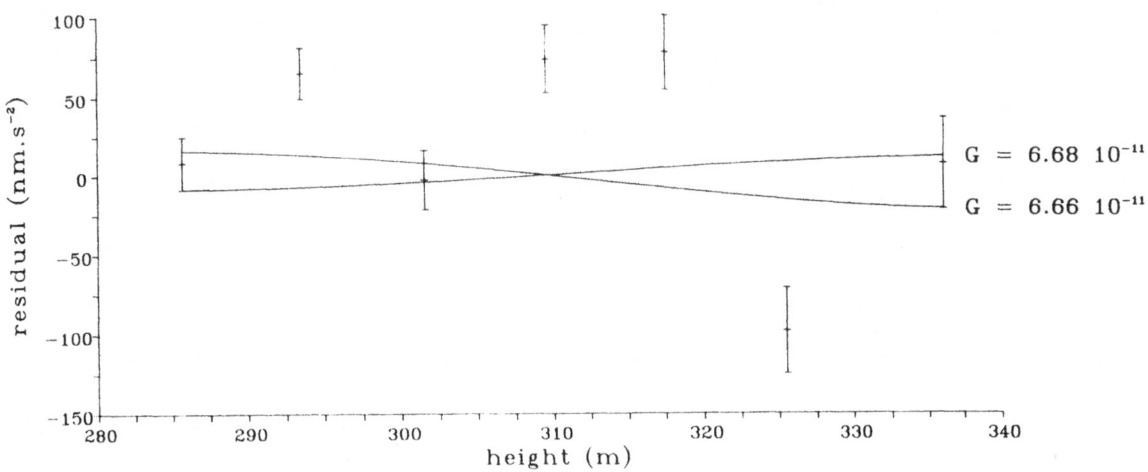

Figure 1: Gravity residuals in Megget Water Tower

the tower shows that the error must lie in computing the effect of the tower itself and, once any error is admitted, doubt is cast on the whole calculation. The experiment will only be convincing if the residuals have the magnitude expected from the observational data and are randomly scattered with respect to height. A probable source of the error in the tower model has now come to light: the present solution involves a uniform density found by regression, whereas the design specifications for steel re-enforcing show a progressively larger proportion of steel towards the base. The attraction of the tower will now be recomputed after modifying the effective density with height.

If a convincing balance of calculated and observed gravity can ultimately be achieved, it will imply that sources at distances ranging from a few centimetres in the tower to many tens of kilometres in the natural topography have all satisfied the inverse square law. The potential precision of better than 1 part in 1000 for the gravitational constant, combined with a demonstrable ability to determine the topographic and subsurface effects, makes the submerged tower form of experiment a very attractive approach for testing Newton's law of gravitation.

Acknowledgements. We wish to thank Lothian Regional Council's Department of Water & Drainage for their collaboration and assistance.

REFERENCES

Heiskanen, W & Moritz, H (1967). *Physical Geodesy,* Freeman, San Francisco.

Hipkin, R G (1988) Bouguer anomalies and the geoid: a reassessment of Stokes' method, *Geoph. J.,* **92**, 53-66.

Hipkin, R G & Hussain, A (1983) Regional gravity anomalies: I Northern Britain, *Rep. Inst. Geol. Sci.,* **82/10**, HMSO London.

Hipkin, R. G., Lagios, E., Lyness, D & Jones, P. (1988) Reference gravity stations on the IGSN71 standard in Britain and Greece. *Geoph. J.,* **92**, 143-148.

Rapp, R. H. & Cruz, J. Y. (1986) Spherical harmonic expansions of the Earth's gravitational potential to degree 360 using 30' mean anomalies, *Rep. Dept. geod. Sci. Surv., Ohio State Univ.,* **376**.

Talwani, M. & Ewing, M. (1960) Rapid computation of the gravitational attraction of three-dimensional bodies of arbitrary shape, *Geophysics,* **25**, 203-225.

LABORATORY METHOD OF CALIBRATING LACOSTE AND ROMBERG MODEL-D GRAVITY METERS.

Jacques Liard
Geological Survey of Canada
Geophysics and Terrain Sciences Branch
Geophysics Division
1 Observatory Crescent
Ottawa, Ontario
Canada K1A 0Y3

INTRODUCTION

A method for the detection and measurement of periodic errors, also known as "circular errors", in the LaCoste & Romberg model D gravity meters has been tested in the laboratories of the Geophysical Division of the Geological Survey of Canada (GSC). This method involves a computer controlled apparatus which performs the measurements automatically as well as the preliminary analysis of the observations. The ground work to establish methodologies and analysis techniques was done on the gravity meter D-28 of the GSC.

The results of the first experiment showed well-defined circular errors having measured amplitudes ranging from 0.1 μGal for the errors of short period to more than 2 μGal (1 μGal = 10 nm/s^2) for the longer periods. The amplitudes were largest for the expected 3.25 mGal period (3250 μGal). Later tests also showed that the same results could be obtained with fewer samples during the automated calibration. Three other instruments were calibrated this way with equally good results.

The presence of large amplitudes at a period shorter than normally quoted in the literature was one unexpected result of this calibration technique. The discovery of an unexpected period in instrument D-28 led to a search for similar periods in the other model D gravity meters.

HISTORICAL TECHNIQUES

The literature generally describes five periods in model D gravity meters, that is, 361.1, 722.2, 1625., 3250, and 6500 CU (Becker et al. 1987). They are all related to the two sets of gears found in the instruments and the 3250 CU period corresponds to one full turn of the micrometer screw.

The main laboratory technique for calibrating model G meters at LaCoste & Romberg is known as the "CloudCroft-Jr" method (Lambert & Liard, 1981; LaCoste & Valliant, 1989). Small repeatable deflections applied to the beam of the instrument are used to map out the response curve of the micrometer screw. However, the CloudCroft-Jr technique is not well suited for measuring circular errors in model D gravity meters since the expected period of 3.25 mGal is far too short for the smaller deflection of 20 mGal.

The technique used by many researchers can be called the staircase method (Becker 1981). This method uses a high precision gravity network in a set of vertical steps previously observed with many model D gravity meters. The maximum gravity difference is usually about 10 mGal and with the help of the instrument RESET screw, the full 200 mGal range can be measured by overlapping sequences.

However, this technique presents a few problems. The gravity meter is constantly handled, and re-leveled from step to step, which leads to a larger standard deviation of the readings. Also, the network itself has to be established with model D gravity meters which will lead to possible biases.

Lately, a few institutions have established another method where the dial of the instrument is calibrated against its own RESET screw. This latest technique as well as the CloudCroft-Jr method reduce the total reading errors as the instrument is not handled repeatedly, nor is it re-leveled from reading to reading, and the 20 mGal deflection, although not well known, is nonetheless a constant.

THE METHOD

In a gravity meter, the beam moves between two capacitive plates and the instrument electronics transform the changing capacitance into a voltage. The output voltage can be used as a method to monitor the beam position which is sampled from the lower stop to the upper stop. Any deviation from a "typical" curve can be taken as the effect of a circular error.

In summary, the method first takes advantage of the RESET screw in order to calibrate the full range of the instrument by taking overlapping sets of samples, and second, changes the "sensitivity" of the instrument so that a greater range of the dial will be covered for any beam displacement between the lower and upper stops. For the purpose of this work, the stop-to-stop range has been increased to about 4000 counter units (CU) instead of the regular 1400 CU. Figure 1 illustrates the apparatus used.

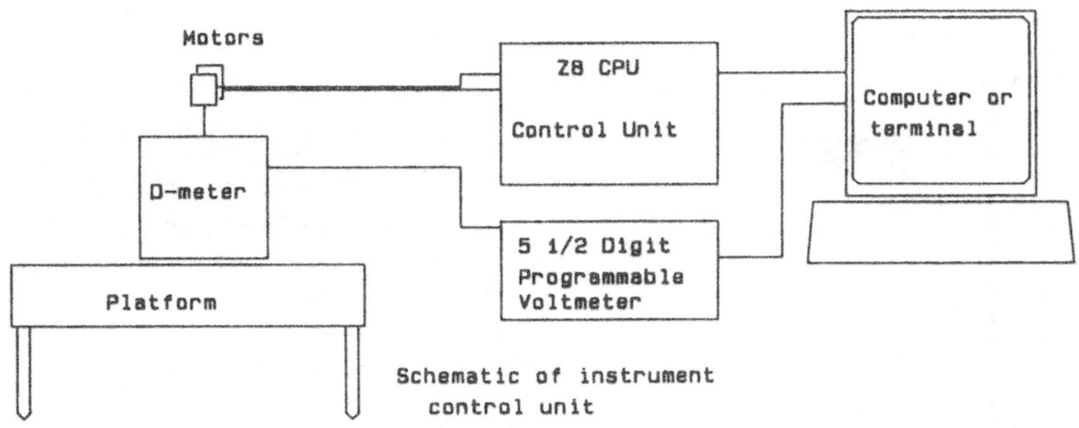

Schematic of instrument
control unit

Figure 1

Each sampled segment generates a set of residuals which can be
plotted with respect to each other. Figure 2 clearly shows that
residuals for one curve are found on subsequent curves for the
same dial positions. Irregularities between individual curves
come mainly from different ground noise levels present during
sampling. Averaging the residuals from consecutive sets further
reduces this noise.

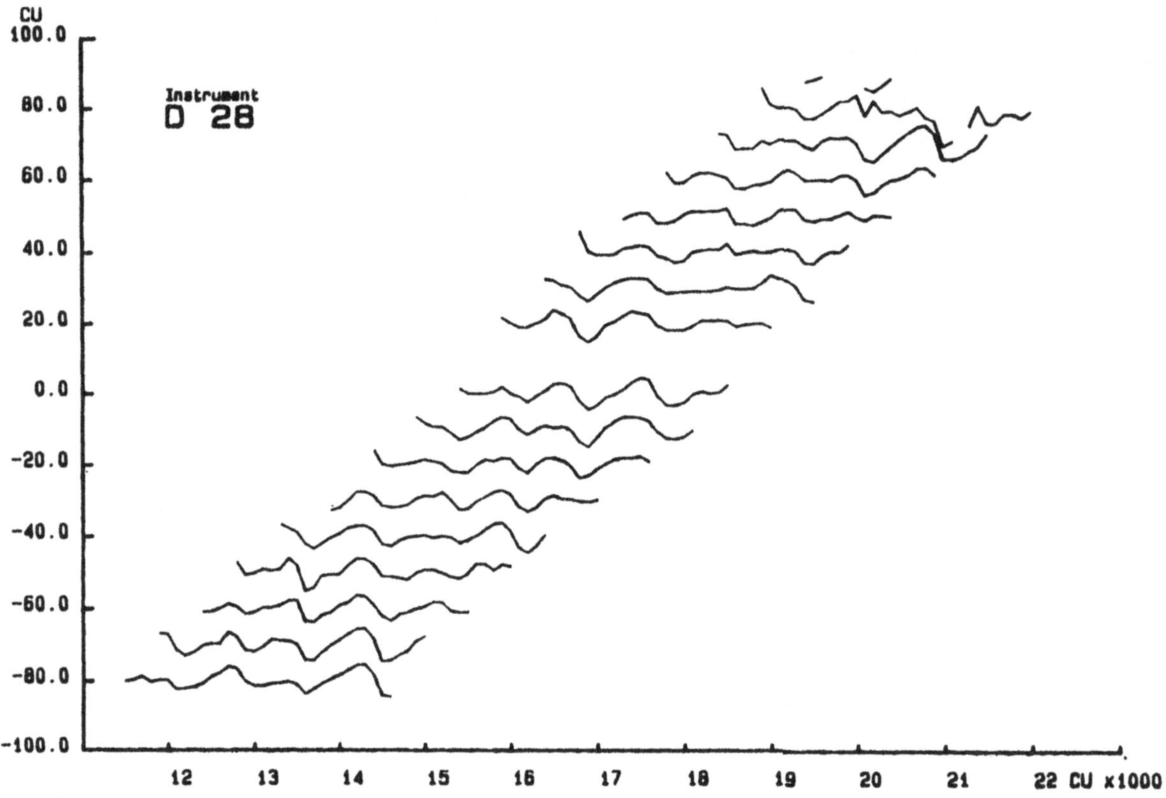

Consecutive sampled segments aligned according to their starting points on the dial

44

A Fourier transform (Figure 3) was applied to the combined
residual curves and a peak at 812.5 CU appeared on the spectra
of the four instruments tested. This peak is unexpectedly large
even though it turns out to be the fourth harmonic of the main
period. A least mean square method was also used to calculate
the phases and amplitudes to confirm and refine the error
estimates of each parameter. Table 1 gives the details of the
amplitudes and phases found for all four instruments with the
least mean square method.

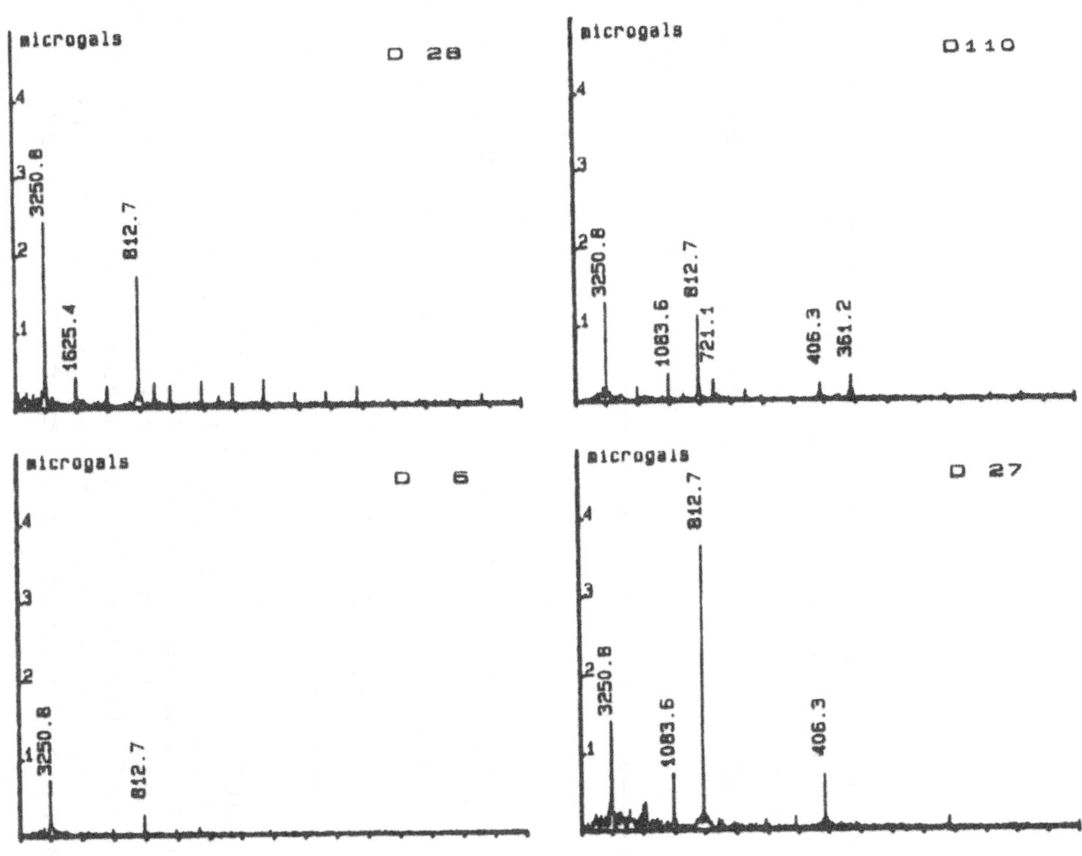

Figure 3

Spectra of the residuals of instruments D-6, D-27, D-28 and D-
110. Significant peaks are identified by their periods in
counter units. (The vertical scale is in microgals but it could
also be in CU.)

Table 1   Results of the automated calibration for four instruments
using the least mean square method.

Periods, amplitudes and phases in Counter Units

Periods	361.1		722.2		812.5		1083.3		1625.		3250.		6500.	
Instr.														
D-6	-	-	-	-	0.3	130	0.1	607	-	-	0.7	3091	-	-
D-27	-	-	0.2	558	3.8	52	0.7	1045	0.4	978	1.4	976	0.2	3508
D-28	0.2	342	0.5	288	1.9	624	0.3	196	0.4	1051	2.6	2494	-	-
D110	0.3	348	0.3	204	1.2	711	0.4	830	0.2	291	1.3	1191	-	-

A graphic representation (Figure 4) of these errors as a
function of one turn of the screw illustrates well the
character of the signals. The circular errors are not as
smooth as expected. A sum of trigonometric functions at the
proper periods and phases will approximate the true character
of the circular errors and will hopefully correct the effects
of such errors on any precise gravity survey network.

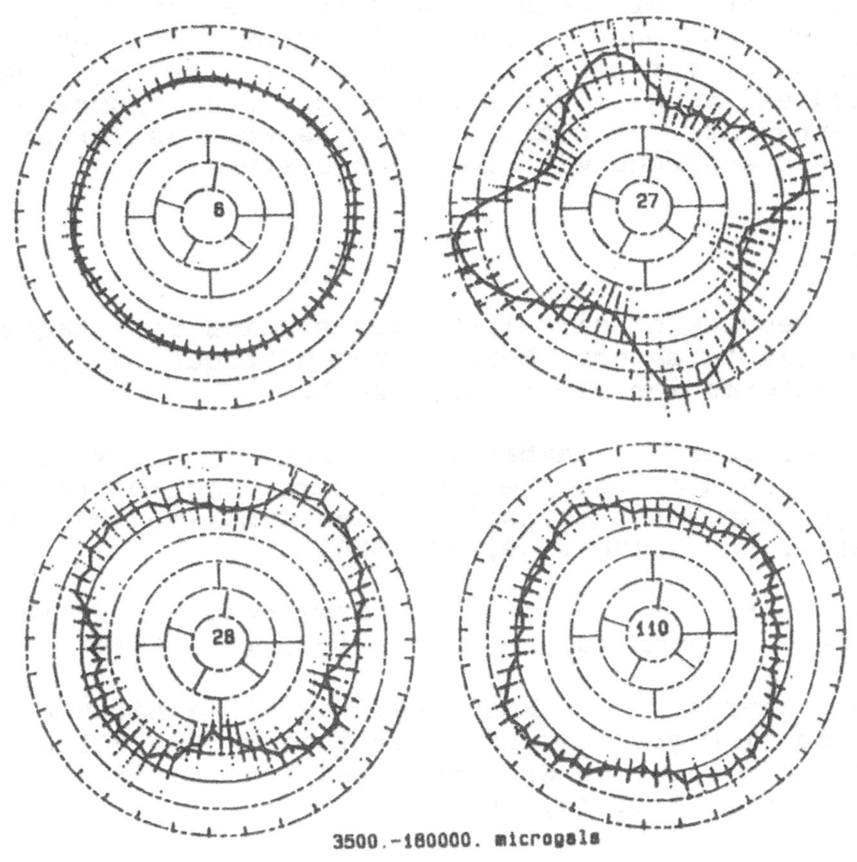

Figure 4

The residuals are plotted on circular diagrams where one full
turn equals 3250 CU and the starting reference is at 0 CU on
the dials of the instruments . Each concentric circle
represents 5 CU and the third circle from the rim is the zero
reference. The radial lines in the inner three circles show
where the 812.5 and 722.2 CU would be. The outer radial lines
indicate intervals of 100 CU.

CONCLUSION

In view of the increasing use of feedback systems being
installed on LaCoste and Romberg instruments, periodic errors
will still have to be accurately measured in an efficient
manner as the range of feedback systems does not exceed some of
the larger periods found in the dial system which seems to be
the main cause of the periodic errors. The knowledge that an
extra period exists will certainly help reduce discrepancies
between instruments for older surveys as well as for gravity
meters not yet equipped with feedback systems.

REFERENCES

Becker, M., Results of circular error studies with LaCoste and
    Romberg gravity meters, Bureau Gravimétrique International,
    Bull. d'Information, 1981, 49, 72-94.

Becker, M., Groten, E., Lambert, A., Liard, J.O., Nakai, S., An
    intercomparison of LaCoste and Romberg Model-D gravimeters:
    results of the International D-meter Campaign 1983, Geophys.
    J. R. astr. Soc., 1987, 89, 499-526.

LaCoste, L. J. B., Valliant, H. D., Gravity meter calibration
    at LaCoste and Romberg, reference in print, 1989.

Lambert, A., Liard, J. O., Nonlinearities in a LaCoste and
    Romberg model-D gravimeters determined by the "Cloudcroft
    Junior" method, Bureau Gravimétrique International, Bull.
    d'Information, 1981, 49, 95-107.

LARGE-SCALE ABSOLUTE GRAVITY CONTROL IN BRAZIL

C. Gemael, O.H.S. Leite, F.A. Rosier
Curso de Pós-Graduação em Ciências Geodésicas
Universidade Federal do Paraná (UFPR)
Curitiba, Paraná, Brazil

and

W. Torge, R.H. Röder, M. Schnüll
Institut für Erdmessung (IfE)
Universität Hannover
D-3000 Hannover, F.R.G

ABSTRACT

Since 1986, IfE Hannover operates the absolute gravimeter JILAG-3. After laboratory and field experiments, hard- and software improvements, as well as comparative measurements at absolute and relative stations, the instrument has been employed for absolute gravity control in fundamental and geodynamic networks in central and northern Europe. At approximately 100 gravity determinations at about 60 different stations an average station precision of a few μgal ($1\ \mu gal = 10\ nms^{-2}$) could be achieved by 500 to 2000 drops. Taking residual instrumental and local effects into account, the average accuracy is estimated to be 10 μgal.

A large-scale control survey has been scheduled and partly performed in 1988/89 in South America, in cooperation with national University Institutes and state agencies. This control system shall cover about 2/3 of South America by establishing more than 20 absolute gravity stations with additional local eccenters. The gravity stations shall partly serve as stations of the *International Absolute Gravity Basestation Network (IAGBN)* proposed by the *International Association of Geodesy (IAG)*, and will also provide gravimeter calibration lines in the countries involved. By connection to the national gravity networks, I.G.S.N.71 in South America will be improved and densified. After an observation campaign in Venezuela in 1988, a project in Brazil has been performed from February to March 1989 in cooperation between IfE and UFPR, covering the country with seven stations. Objectives of the project and first results of the measurements are presented and discussed in the paper.

INTRODUCTION

In February/March 1989 gravity observations have been carried out in Brazil with the absolute gravimeter JILAG-3 of the Institut für Erdmessung (IfE). These observations are part of an absolute gravimetry program in Latin-America. Up to now three stations in *Argentina*, seven in *Brazil*, two in *Uruguay*, and six in *Venezuela* have been determined. By these measurements an absolute gravity control system is established in the eastern and central part of South-America which primarily serves for the objectives defined by the *International Association of Geodesy (IAG)*. It includes three stations of the proposed

International Absolute Gravity Basestation Network (IAGBN) (Boedecker and Fritzer 1986), one of which is located near Brasilia/Brazil. Simultaneously the IAGBN is densified on regional level, leading to the strengthening and homogenization of the existing base-networks and the stabilization of geodynamic networks in the central and Venezuelan Andes. In addition centering measurements to stations at floor level and to outdoor stations have been performed. For these connections two LaCoste-Romberg gravimeters with electrostatic feedback (type SRW, designed at IfE) have been employed.

The field work in Brazil has been performed by O.H.S. Leite and F.A. Rosier from UFPR and R.H. Röder, M. Schnüll and D. Beening from IfE.

GRAVITY MEASUREMENTS IN BRAZIL

Absolute gravity measurements

At present time there exist about a dozen transportable absolute gravimeters worldwide. Since January 1986 one of these instruments (JILAG-3) is operated by IfE. It is one of six Faller-type instruments, constructed by Prof. J.E. Faller at the Joint Institute for Laboratory Astrophysics (JILA), National Bureau of Standards and University of Colorado, which use simultaneous measurements of length and time (laser interferometry) during the free fall of a mass, to determine gravity (Faller et al. 1983). The instrument is highly automated and can be easily operated by only two persons. The accuracy of a gravity value derived from a large number of drops is estimated to reach the order of a few μgal (Niebauer 1987). From repeated measurements on seven stations in the Federal Republic of Germany a r.m.s. discrepancy of $\pm 9 \mu gal$ was found for JILAG-3, which indicates the existence of residual effects probably due to varying instrumental and station conditions (Torge et al. 1987). Comparisons with gravity values determined by other instruments and by combined absolute and relative measurements showed a systematic discrepancy of appr. $-25\,\mu gal$ (JILAG-3 - other result). Early 1988, this discrepancy could be remarkably reduced by a re-adjustment of the instrument's interferometer base, resulting in a correction of $+22\mu gal$ to the JILAG-3 results obtained before February 1988 (Torge et al. 1989). From IfE experiences in the last three years, the present accuracy of JILAG-3 is estimated to be $\pm 10\,\mu gal$ or better.

The JILAG-3 absolute gravimetry equipment consists of the absolute gravimeter inside its original transportation boxes, two LaCoste-Romberg gravimeters, related spare parts and tools. It was packed into 7 aluminium containers and sent to Brazil by regular air freight. The weight is 740 kg including the packing material. After the arrival in Brazil some electronic components were found to have got loosen, though the instrument was protected by a great amount of rubber foam. From the first station to the second one the equipment was transported by air, and in the following on the loading area of an open 1-t truck, covered with a plastic awning, without using the containers. Because of limited space and missing possibilities for fixing, the boxes housing the dropping chamber and the superspring could not be transported in upright position during air and car travel, as is done usually. To maintain the vacuum, the ion pump was connected to a 63-Ah dryfit battery which keeps power supply for five days at least. Small repairs were necessary at nearly every station, mainly due to the rough transportation conditions. After having observed five stations in Brazil the steel belt, which drives the drag free chamber had to be exchanged. Pumping down the dropping chamber again caused a three days delay.

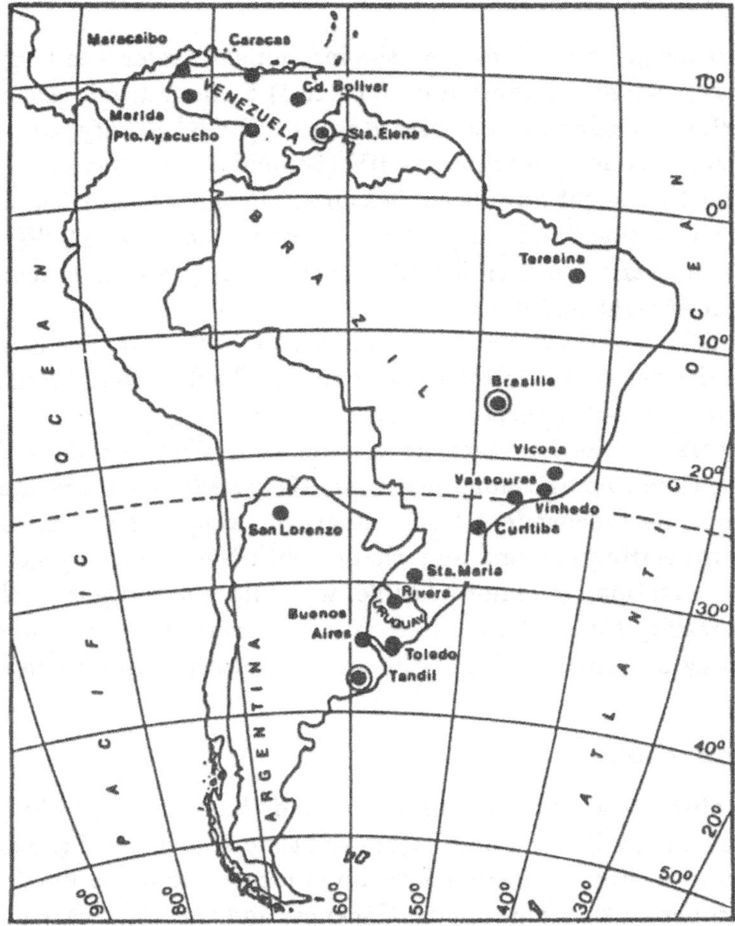

Fig. 1. Location of absolute gravity stations in Brazil.

Measurements started in Teresina on February 14, 1989. In the following four weeks the stations Brasilia, Viçosa, Vassouras, Vinhedo, Curitiba, and Sta. Maria have been occupied (see Fig. 1.). Between 1500 and 3000 (altogether 16200) drops have been performed at these stations in sets of 300 or 600. From every drop, 155 time-distance pairs were fitted to a parabola yielding one gravity value. Criterion for the acceptance of these values was the scatter of the measurement (r.m.s. distance residuals). According to our experiences the result of a drop contains a gross error, if the scatter exeeds 10 nm. These drops are rejected. An earth tide correction (constant amplitude factor 1.164, phase shift zero) is applied to every drop, whereas corrections for polar motion, deviation of actual air pressure from normal air pressure and a temperature dependent correction for the laser wavelenght are applied to the mean value of a set of measurements. The standard deviations for the average of a single accepted drop on a station varied between 32 and $97\mu gal$, depending on the station conditions, with local ground vibrations having a large influence. They were mainly caused by air conditioning, but could not be avoided because of high temperatures up to $30°C$ during the working period.

Relative gravity measurements

For the reduction of the gravity values from the instrument's reference height (about 81 cm depending on the respective installation of JILAG-3) to floor level, relative gravity measurements were performed using two LaCoste-Romberg (LCR) gravity meters equipped with electrostatic feedback systems developed at IfE (Schnüll et al. 1984). The differences between floor level and 1 m height have been measured ten times each with both instruments. The accuracy of the resulting "vertical gradient" is estimated to be $\pm 2 \frac{\mu gal}{m}$. In Brasilia the linearity of the gravity variation with height was checked by also measuring the difference between the reference height and floor.

Nine eccenter stations were connected to the absolute stations, five of them being sites of the national gravity networks. In three cases the feedback systems could not be used as the gravity differences exceeded their measuring ranges.

Calibration factors for the LCR-gravity meters were determined in the "Gravimeter Calibration System Hannover" (Kanngieser et al. 1983). While calibration parameters for the feedback systems are independent from the gravity range, the parameters for the sole LaCoste-Romberg measuring systems may change with the gravity range. As relative ties between our absolute stations were not possible, we could not verify the calibration parameter values in the gravity range of Brazil. Periodical errors could not be corrected. Thus a few μgal error may be introduced into the eccenter stations due to calibration errors.

Results and comparisons

The results of the absolute gravity measurements and the centering to floor level are given in Table 1. For the stations Brasilia and Viçosa the time sequences of the drops and related histograms are shown in Fig. 2. Station Brasilia is located some 15 km from the city in a natural park on a concrete floor with 35 cm thickness and therefore very quiet, whereas the observations in Viçosa were disturbed by heavy noise from air conditioning and traffic on a nearby road.

Table 1. Results of absolute gravity measurements in Brazil 1989.

Station	No.of drops	$g_{ref.height}$ mgal	Ref.height m	$\frac{\Delta g}{\Delta h}$ $\frac{mgal}{m}$	$g_{floor\ level}$ mgal
110/112 Teresina	1980	978016.0969	0.802	0.3039	978016.341
120/122 Brasilia	1500	978048.5898	0.808	0.2478	978048.790
130/132 Viçosa	3000	978459.9933	0.806	0.2869	978460.224
140/142 Vassouras	2700	978637.3257	0.813	0.3055	978637.574
150/152 Vinhedo	1980	978563.4745	0.803	0.3677	978563.770
160/162 Curitiba	2940	978760.1239	0.811	0.3120	978760.377
170/172 Sta. Maria	2100	979261.3790	0.807	0.3032	979261.624

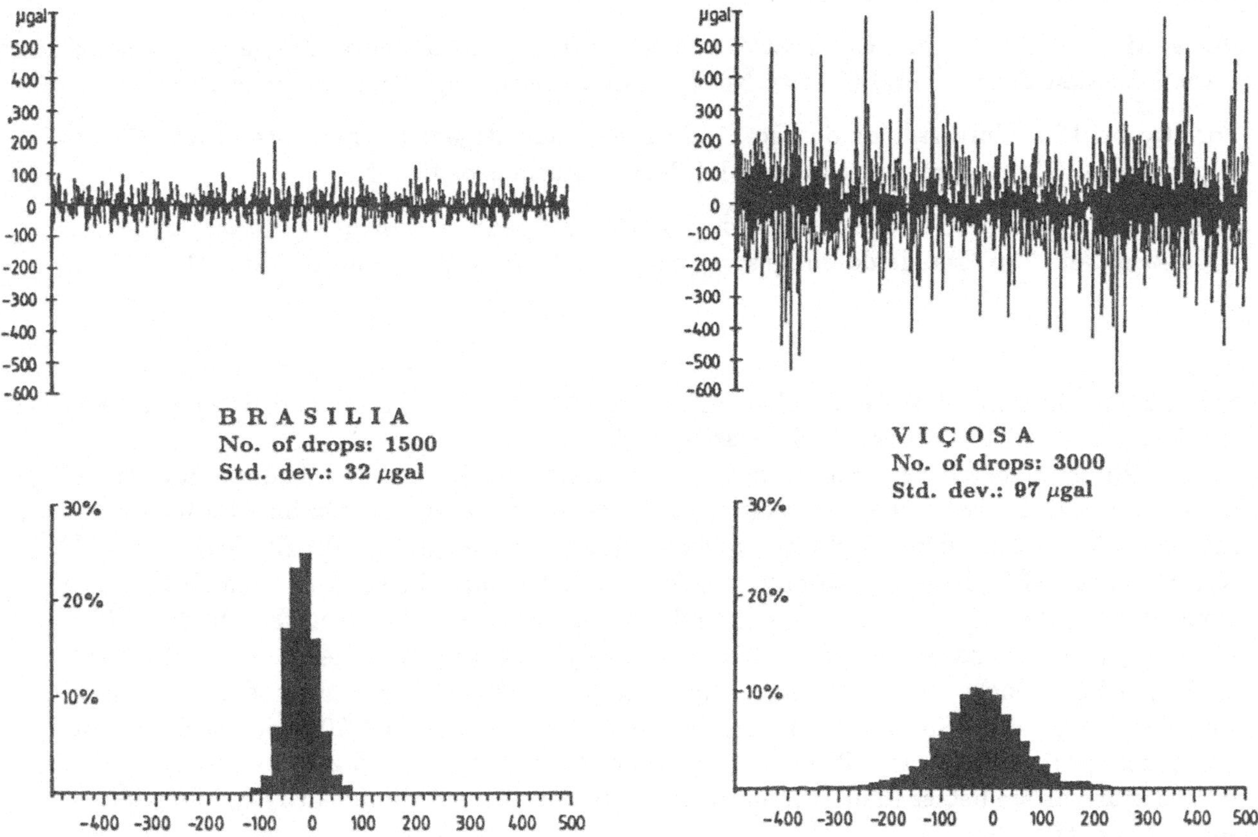

Fig. 2. Time sequences and histograms of absolute gravity observations in Brasilia and Viçosa.

Table 2. contains the gravity values for the eccenter stations and already existing values g_{net} from local gravity nets as well as the discrepancies between the two determinations.

Table 2. Eccenter stations and comparison.

Station	$g_{JILAG-3}$ mgal	g_{net} mgal	$g_{JILAG-3} - g_{net}$ mgal
111 Teresina	978.016175	—	—
113 Teresina B	978017.373	978017.41	−0.037
123 Brasilia C	978088.356	978088.42	−0.064
133 Viçosa	978459.469	—	—
143 Vassouras B	978637.883	978637.93	−0.048
145 Vassouras	978637.827	—	—
153 Campinas C	978602.389	978602.396	−0.007
163 Curitiba	978761.062	—	—
173 Sta. Maria	979238.195	979238.225	−0.030

Sources for the existing values are:

For stations 113, 123, 143: Observatório Nacional do Rio de Janeiro 1987, adjustment of the Brazilian Fundamental Gravity Net (BFGN). Constraints: 16 stations of IGSN71.

For station 153: Universidade de São Paulo 1989,. adjustment of the State of São Paulo gravity net. Constraints: 1 station of IGSN71, 5 stations of BFGN.

For station 173: Universidade Federal do Paraná 1979, adjustment of the States of Santa Catarina and Rio Grande do Sul gravity net. Constraints: 2 stations of IGSN71.

Standard deviations for the g_{net}-values of 153 and 173 are 0.019 mgal and 0.017 mgal respectively. Those of the other stations are not given.

The adjustment of the absolute data in the single stations gave standard deviations between 1 and 2 μgal (internal precision) for the mean values. A combined adjustment included all absolute values and relative measurements performed in Brazil, Uruguay, and Argentina in 1989. Here the absolute values have been introduced with standard deviations of $\pm 10\,\mu gal$. The accuracies for the relative measurements between the absolute sites and their eccenters have been estimated from long-time experiences. As a priori standard deviation for a single tie we have introduced: 5 μgal (feedback gravimeter, transportation by hand), 10 μgal (feedback-gravimeter, transportation by car), and 20 μgal (feedback not used, transportation by car). The resulting accuracies of the adjusted gravity values of connected stations are better than 12 μgal, so they are not affected significantly by the relative measurements.

Even if taking the uncertainties of the station values into account there is an indication, that the level of the Brazilian gravity networks is some 10 μgal higher than that derived from the 1989 results. This problem has to be investigated in more detail after more network stations have been tied to the absolute sites.

CONCLUSIONS

Through the absolute gravity project "Brazil 1989", seven absolute gravity stations have been established within four weeks. These stations are located mainly in the eastern part of the country, and they include a station near the capital Brasil ia. The average accuracy of the gravity values is estimated to be $\pm 10\mu gal$, by local relative connections existing gravity networks have been tied.

The results of the gravity project "Brazil 1989" thus

- contribute to the establishment of IAGBN,

- serve as a gravimeter calibration line, with appr. 1250 mgal range,

- provide reference stations for the national gravity network.

It is intended to continue this fundamental work by

- establishing absolute gravity stations also in the central and western parts of Brazil,

- strengthening the connections between absolute stations and the national network, and

- analyzing and eventually improving the fundamental gravimetric network by a combined adjustment.

Acknowledgements. The authors from UFPR are grateful to the IBGE (Instituto Brasileiro de Geografia e Estatística), Petrobrás (Petrobrás S/A), Varig (Viação Aérea Riograndense), the Universities: UFPR, UFPI, UNB, UFV, USP, UFSM and the Observatório Nacional which have supported the research in Brazil. The authors from IfE are grateful to the German Research Society (Deutsche Forschungsgemeinschaft) for providing the absolute gravimeter and a research grant (To 46/29-1) which allowed the investigations presented here.

REFERENCES

Boedecker, G., and Fritzer, Th. (1986). International Absolute Gravity Basestation Network. IAG-SSG 3.87 Status Report March 1986, Veröff. Bayer. Komm. f. d. Int. Erdmessung d. Bayer. Akad. d. Wiss., Astron. - Geod. Arb., Nr. 47, München.

Faller, J.E., Guo, Y.G., Gschwind, J., Niebauer, T.M., Rinker, R.C., and Xue, J. (1983). The JILA portable absolute gravity apparatus. Bur. Grav. Int., Bull. d'Inf., **53**, 87-97.

Kanngieser, E., Kummer, K., Torge, W., and Wenzel, H.-G. (1983). Das Gravimeter-Eichsystem Hannover. Wiss. Arb. Fachber. Verm.wesen, Univ. Hannover, Nr. 120.

Niebauer, T.M. (1987). New absolute gravity instruments for physics and geophysics. Ph. D. Thesis, Univ. of Colorado, Boulder.

Schnüll, M., Röder, R.H., and Wenzel, H.-G. (1984). An improved electronic feedback for LaCoste-Romberg gravity meters. Bur. Grav. Int., Bull. d'Inf., **55**, 27-36.

Torge, W., Röder, R.H., Schnüll, M., and Faller, J.E. (1987). First results with the transportable absolute gravimeter JILAG-3. Bull. Géodésique, **61**, 161-176.

Torge, W., Röder, R.H. and Schnüll, M. (1989). Correction of a systematic error of gravity measurements with JILAG-3. Bur. Grav. Int., Bull. d'Information, in press.

A SPACEBORNE GRAVITY GRADIOMETER FOR THE NINETIES

by A. Bernard and P. Touboul
Office National d'Etudes et de Recherches Aérospatiales,
BP 72, 92322 Châtillon Cedex, France

INTRODUCTION

To improve the global accuracy of Earth Gravity field models, ground techniques and conventional satellite methods cannot be used much further. Satellite Gravity Gradiometry (SGG) is the most promising way to explore gravity variations in the resolution range of 100 km over lands as well as over oceans.

For the nineties, a great step forward is expected with the european ARISTOTELES mission aiming at a resolution of 10^{-2} Eötvös at an altitude of 200 km during six months.

The gravity gradient components will be determined through differential measurements between the GRADIO electrostatic ultrasensitive accelerometers which are being developed in ONERA.

Designed to achieve a resolution of 10^{-11} ms$^{-2}/\sqrt{Hz}$ for accelerations less than 10^{-4} ms^{-2}, the accelerometer configuration is compatible with an operation on ground. During 1988, a first laboratory model has been manufactured and tested.

A plane gradiometer composed of four accelerometers and of a calibration system will be accommodated on board a non drag free satellite. In spite of the jointly optimized designs of the spacecraft and of the gradiometer, the data processing will have to extract the gravity variations from the much larger disturbances due to drag or angular motion.

SPACEBORNE MICRO ACCELEROMETRY

The gravitational field is derived from the expression of the potential u: $\vec{g} = \nabla \vec{u}$ and the second derivative leads to the gravity gradient tensor [T] defined by:

$$T_{ij} = \frac{\partial^2 u}{\partial x_i \, \partial x_j}$$

The spatial variations of this tensor components can be measured by comparing the outputs of different accelerometers on board a satellite. Because it is symmetric and traceless, only five components are independent.

Let us consider with the figure 1 one accelerometer on board a satellite. The accelerometric measurement rests on the expression of the proof-mass dynamics:

$$\vec{F_i} + m_i\,\vec{g}(B_i) = m_i\,\vec{\Gamma}(B_i) \quad \text{(see notations of figure 1)}$$

The accelerometer output $\vec{\gamma_i}$ obtained from the measure of the applied forces $\vec{F_i}$ can be expressed as the difference of the acceleration of B_i (with respect to the inertial reference frame) and of the gravity field $\vec{g}(B_i)$ at point B_i :

$$\vec{\gamma_i} = \frac{\vec{F_i}}{m_i} = \vec{\Gamma}(B_i) - \vec{g}(B_i).$$

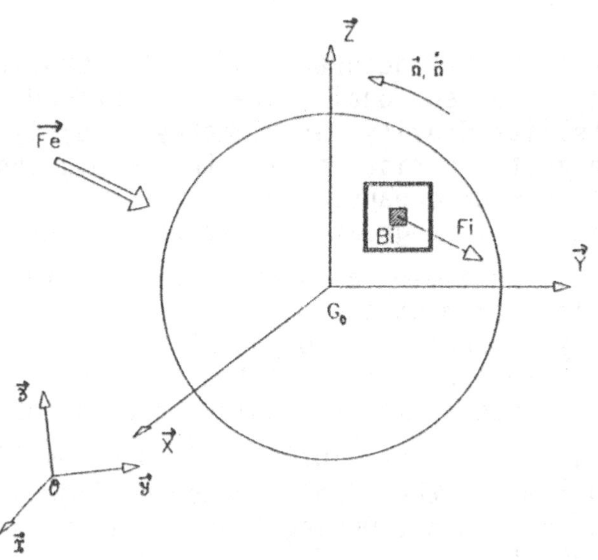

Fig. 1. Micro-accelerometry.

<u>Notations:</u>

M, G_{0S}	: mass, centre of mass of the satellite main body S_c,
O_i	: centre of the accelerometer cage (solid to the satellite main body),
m_i , B_i	: mass and centre of mass of the accelerometer proof-mass,
G_0	: centre of mass of the composite, S, constituted by the satellite main body and the masses m_i ,
O, $\vec{x}$, $\vec{y}$, $\vec{z}$	: inertial reference frame,
G_0, X, Y, Z	: satellite reference frame : linked in translation to the centre of mass of the composite and linked in orientation to the satellite main body,
$\vec{\Omega}$, $\dot{\vec{\Omega}}$	: angular velocity and acceleration of the satellite main body with respect to the inertial frame,
$\vec{F}_{ext}$	: external forces acting on the satellite (non gravitational),

$\vec{F}_i$: force acting on the accelerometer proof-mass (non gravitational).

By developing the acceleration $\vec{\Gamma}(B_i)$ according to the acceleration of the centre of mass G_0 of the composite body (satellite main body and other moving masses m_i) and according to the attitude motions of the satellite, $\overrightarrow{\gamma_i}$ can be written as the sum of five terms:

$$\overrightarrow{\gamma_i} = \frac{\overrightarrow{F_{ext}}}{(M + \Sigma\, m_i)} \qquad \text{External acceleration (drag, ...)} \qquad (1)$$

$$+ \int_S \vec{g}_{(m)}\, dm - \overrightarrow{g_{(B_i)}} \qquad \text{Gravity gradient}$$

$$+ \frac{M}{(M + \Sigma\, m_i)} \left\{ \vec{\Omega} \wedge \left(\vec{\Omega} \wedge \overrightarrow{G_{0S}O_i} \right) + \overset{o}{\vec{\Omega}} \wedge \overrightarrow{G_{0S}O_i} \right\} \begin{array}{l}\text{Satellite}\\ \text{angular}\\ \text{motion}\end{array}$$

$$+ \frac{M}{(M + \Sigma\, m_i)} \left\{ 2\vec{\Omega} \wedge \overset{o}{\overrightarrow{G_{0S}O_i}} + \overset{oo}{\overrightarrow{G_{0S}O_i}} \right\} \begin{array}{l}\text{Change of satellite}\\ \text{geometry and mass}\\ \text{distribution}\end{array}$$

$$+ \frac{M}{(M + \Sigma\, m_i)} \left\{ \vec{\Omega} \wedge \left(\vec{\Omega} \wedge \overrightarrow{O_iB_i} \right) + \overset{o}{\vec{\Omega}} \wedge \overrightarrow{O_iB_i} + 2\vec{\Omega} \wedge \overset{o}{\overrightarrow{O_iB_i}} + \overset{oo}{\overrightarrow{O_iB_i}} \right\}$$

$$\text{Proof-mass relative motion}$$

The proof-mass of the accelerometer is maintained motionless with respect to the cage by the electrostatic suspension with six servo-loop controls. Thus the last term depending on $\overrightarrow{O_iB_i}$, $\overset{o}{\overrightarrow{O_i\,B_i}}$ and $\overset{oo}{\overrightarrow{O_iB_i}}$ will be kept very close to zero especially inside the bandwidth of the transfer function of the accelerometers.

In order to extract the gravity gradient from the differential measurements of two accelerometers of the gradiometer it is necessary to reject the external perturbing terms coming from the drag, the radiation pressures, the satellite angular motion, the change of satellite mass distribution, ...

Besides the outstanding resolution to be achieved for the accelerometers, the rejection of these perturbing terms requires:
- a good linearity of the accelerometers,
- low couplings between axes,
- very accurate matchings of scale factors and of alignments.

For the ARISTOTELES mission (Bernard, 1988), the choice of a non drag free satellite for a low orbit Earth gravity mission leads to accommodate the plane gradiometer, instead of a full tensor instrument, at the centre of mass of the satellite in the plane normal to the drag (see Fig. 2).

   At the altitude of 200 km, the DC level of the drag acceleration that
can reach $10^{-4}\mathrm{ms}^{-2}$ at the end of the mission. Because of this high
level, the resolution of 10^{-2} Eötvös cannot be obtained along the
velocity axis $\vec{x}$.

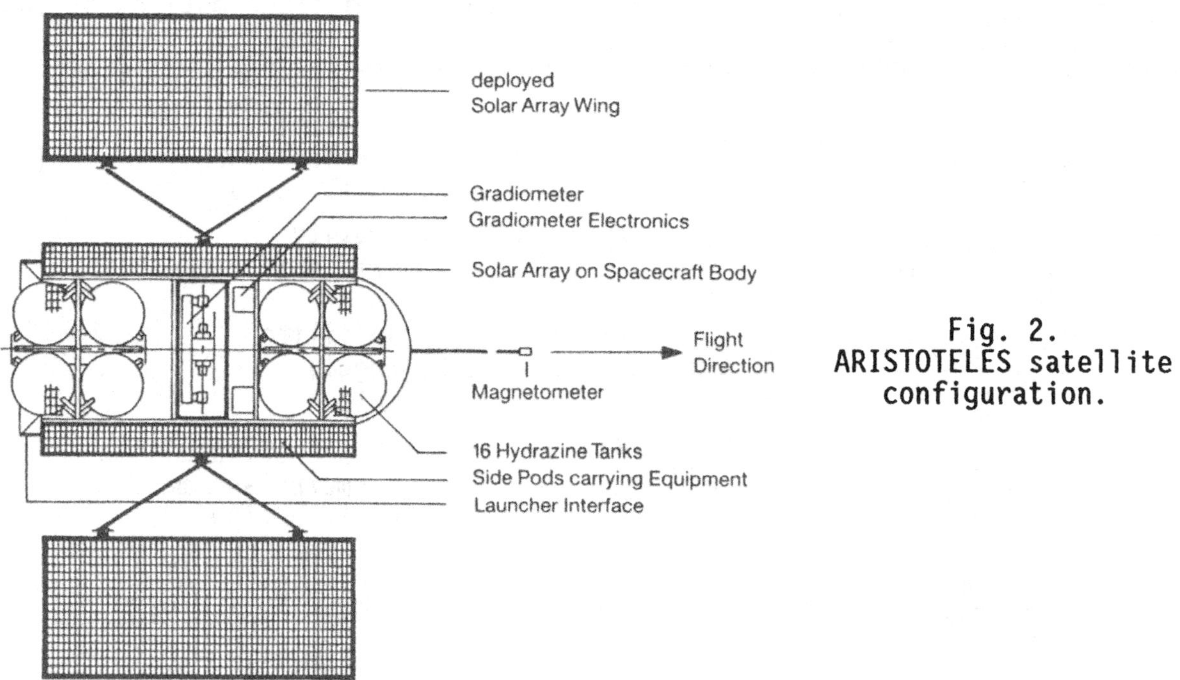

Fig. 2.
ARISTOTELES satellite
configuration.

   The main objective of the plane gradiometer as accommodated in the
ARISTOTELES satellite is to determine the variations of the tensor
components T_{yy} and T_{zz} with a resolution of 10^{-2} Eötvös in the
measurement bandwidth corresponding to periods between 200 s and the 6 s
integrating time.

   The acceleration $\vec{\gamma_i}$ -see expression (1)- measured by each
accelerometer is affected by the uncertainties of bias, scale factor and
orientation of each accelerometer. This leads us to express the actual
measure Γ_i as:

$$\vec{\Gamma_i} = k_0 + k_1\,\vec{\gamma_i} + k_{\ell m}\,\vec{\gamma_i} + k'_{\ell m}\,\vec{\gamma_i} + k_2\,\gamma_i \otimes \gamma_i + noise$$

with: k_0 : bias vector,
k_1 : scale factor matrix,
$k_{\ell m}$: misorientation matrix (with respect to instrument frame),
$k'_{\ell m}$: coupling matrix,
k_2 : quadratic non linearity matrix.
$k_{\ell m}$ is an antisymmetric matrix that represents the orientations of the
sensitive axes of the accelerometer with respect to the instrument
frame. $k'_{\ell m}$ is a symmetric matrix that represents the coupling between
the three axes of the GRADIO accelerometer. This coupling is due to the

defects of geometry of the proof-mass that can be reduced, thanks to an optical grinding, to less than 10^{-5} rad i.e. coupling less than 10^{-5}.

In the linear combinations of the accelerometer outputs, the measurements of the differences of gravity field between the accelerometers are disturbed by the limited rejections of the four other acceleration terms.

The table 1 summarizes the main instrument requirements that are derived in order to limit the disturbances due to drag, angular velocity and acceleration of the satellite to less than 5×10^{-12} ms^{-2} (i.e. 5×10^{-3} Eötvös with an instrument of 1 m baseline).

PERTURBATION SOURCE	INSTRUMENT REQUIREMENTS
1 - DRAG DC along $\vec{x}$* : 10^{-4}ms^{-2}	Stability of $\vec{y_A}$ (or $\vec{z_A}$)** orientation with respect to $\vec{x}$: 5×10^{-8}rad
2 - DRAG DC along $\vec{y}$ (or $\vec{z}$) : $< 5 \times 10^{-6}$ms^{-2}	Stability of scale factor of $\vec{y_A}$ (or $\vec{z_A}$): 10^{-6} Stability of orientation with respect to $\vec{z}$ (or $\vec{y}$): 10^{-6}rad
3 - DRAG variations (along $\vec{x}$) : $< 5 \times 10^{-6}$ms^{-2} reduced by flaps to : $< 5 \times 10^{-7}$ms^{-2}	Alignment with respect to the plane normal to $\vec{x}$: 10^{-5}rad
4 - DRAG variations along $\vec{y}$ (or $\vec{z}$): $< 5 \times 10^{-6}$ms^{-2}	Scale factor matching of $\vec{y_A}$ (or $\vec{z_A}$) : 10^{-5} Alignment with respect to the plane normal to $\vec{y}$ (or $\vec{z}$) : 10^{-5}rad
5 - Angular velocity : 10^{-6}rad/s	T_{zz} : estimation of the variation of the angular velocity about $\vec{y}$ to 2.5×10^{-9} rad/s
6 - Angular acceleration : 10^{-7}rad/s^2	Alignments of the sensitive axes with respect to the directions defined by the accelerometer centres : 5×10^{-5} rad

* $\vec{x}$, $\vec{y}$, $\vec{z}$ satellite axes

** $\vec{x_A}$, $\vec{y_A}$, $\vec{z_A}$ accelerometer sensitive axes

Table 1. Instrument requirements derived from each perturbation source to limit the effects to less than 5×10^{-12} ms^{-2} i.e. 5×10^{-3} Eötvös for a 1 m gradiometer.

The calibration of the accelerometers is required during the mission to estimate the scale factors and the alignments of the accelerometer sensitive axes, and even their quadratic non linearity. Then the gradiometer will be composed of a very stiff plane structure exhibiting high geometrical and thermal stabilities. This structure will carry, at each corner, one GRADIO accelerometer and at its centre a dedicated calibration system. This mechanism is composed of two pairs of symmetrical unbalanced wheels rotating at two different angular

frequencies. When two by two in phase, these wheels will apply to the
satellite, centred with respect to the instrument, sine wave calibrating
accelerations in translation; in opposite phase, they would apply
angular accelerations. By exploiting the synchronous demodulations of
the accelerometer outputs at the angular frequencies of the wheels or at
twice these frequencies, most of the accelerometer characteristics can
be estimated.

The filtering out of the angular acceleration in the gradiometric
measurement is based on the characteristics of symmetry and
anti-symmetry of respectively the angular acceleration matrix and of the
gravity gradient tensor. For instance $\left(\vec{\Gamma_i} - \vec{\Gamma_j}\right) \cdot \overrightarrow{O_i\,O_j}$ is independent
of $\overset{0}{\vec{\Omega}}$ in the case of a perfectly aligned instrument. Moreover, the
rejection of $\overset{0}{\vec{\Omega}}$ requires to achieve a fine positioning of the
accelerometers on the instrument structure. The directions defined by
the accelerometer centres $\overrightarrow{(Og)}$ must be aligned with the sensitive axes
$\left(\vec{\Gamma_i}, \vec{\Gamma_j}\right)$. Because the angular acceleration of the satellite will be
controlled to less than 10^{-7} rad·s^{-2}, the two matrices $[\dot{\Omega}]$ and $[T]$ must
be decorellated with a ratio greater than 5×10^{-5}.
The angular velocity terms cannot be rejected in the same way
(Bernard, Touboul, 1987). The Earth pointing of the ARISTOTELES
satellite must be accurately stabilized to limit the satellite angular
velocity variations to less than 10^{-6} rad/s as so as the angular
acceleration to less than 10^{-7} rad/s^2. In these conditions, the gravity
gradient component measurement T_{yy} is not affected at first order by the
$[\Omega^2]$ term variations. On the contrary, the measures of T_{zz} are disturbed
by the product $2\Omega_0\delta\Omega_y$ where Ω_0 is the orbital angular velocity about $\vec{y}$
and $\delta\Omega_y$ the fluctuations of Ω_0 in the measurement bandwidth. Then the
resolution of 10^{-2} Eötvös requires to estimate the fluctuations $\delta\Omega_y$ with
an accuracy of at least 2.5×10^{-9} rad/s. This should be achieved by
using the informations provided by the GRADIO accelerometers themselves.

GRADIO ACCELEROMETER

The principle of operation of the accelerometer rests on the measurement
of the force that is necessary to maintain a proof-mass at the centre of
a cage. This force is provided by a three-axis electrostatic suspension.
The GRADIO accelerometer takes advantage of the low levels of
acceleration to undergo (less than 10^{-4} ms^{-2}) for the optimization of
the performances and a solid parallelepipedic proof-mass is used for
such a space accelerometer.
The electrostatic forces remain normal to the equipotential faces of

the conductive proof-mass and thus these directions determine the sensitive axes.

The electrostatic suspension of this proof-mass is achieved by six servocontrol channels, 3 rotations, and 3 translations, acting separately along the three axes of the accelerometer.

The mechanics of the GRADIO accelerometer shown in Fig. 3 is constituted by the proof-mass surrounded by a system of electrodes used at the same time for the generation of the electrostatic suspending forces and, for the capacitive sensing of the proof-mass position.

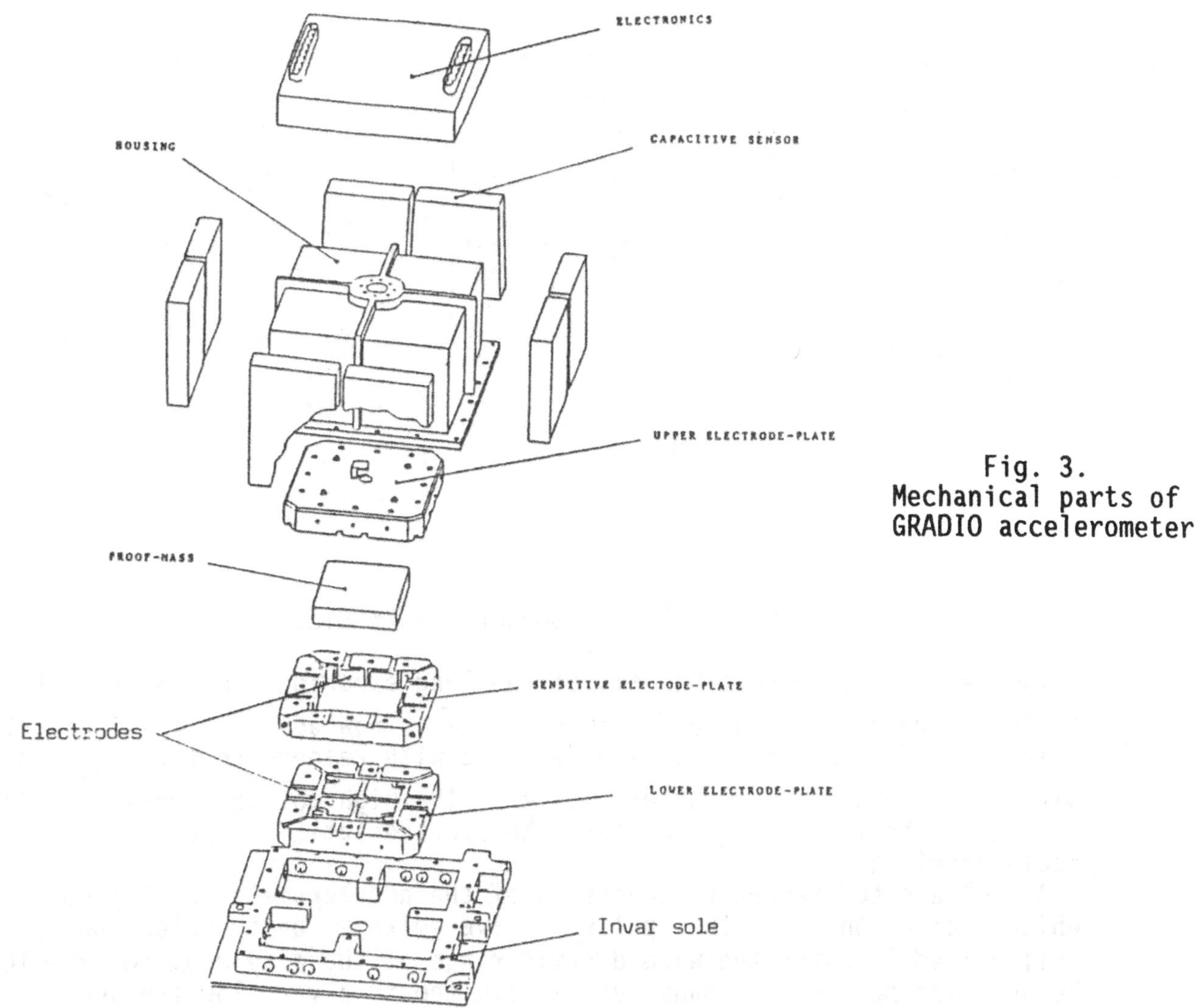

Fig. 3.
Mechanical parts of
GRADIO accelerometer

A platinum-Rhodium alloy is preferred for the proof-mass material because of its high density 20×10^3 kg/m^3, its relatively low coefficient of thermal expansion 8.8×10^{-6}/°C and its low magnetic susceptibility 3×10^{-4} m^{-3}. Furthermore, this alloy experienced with the proof-mass of the CACTUS accelerometer, offers good possibilities for accurate machining and grinding.

The electrodes are obtained by ultra-sonic machining of three plates

made of Ultra Low Expansion Titanium Ceramica. These plates are gold
coated.

The sensitive ring plate is pressed between the two carrier plates and
its thickness sets the gaps between proof-mass and carrier electrodes to
30 μm. The gap between the proof-mass and the sensitive electrodes is
fixed to 300 μm to insure the expected resolution of the accelerometer
to better than 5×10^{-12} ms^{-2} for a full scale range of 10^{-4} ms^{-2} and an
integrating time of 6 s.

The Fig. 4 presents the schema of the electronics.

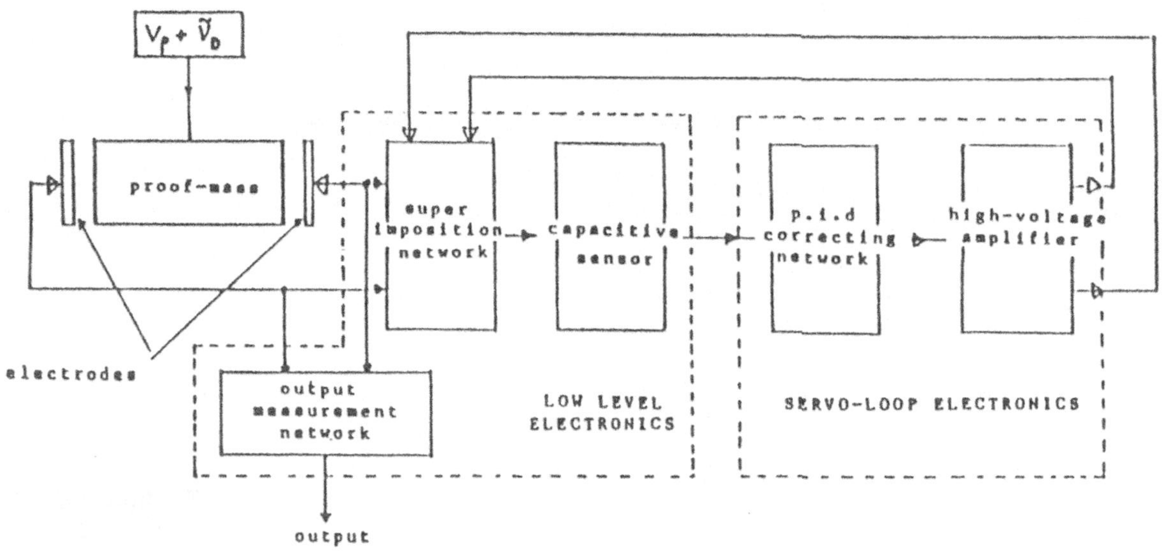

Fig. 4. Accelerometer electronics.

The capacitive sensors benefit of the long experience of ONERA in that
field. A resolution of a few 10^{-7} pF/$\sqrt{Hz}$ is now achievable allowing to
maintain the proof-mass in steady state with respect to the electrodes
with a resolution of about 0.1 Å.   The high voltage amplifier will
generate the 150 V required for the electrostatic suspension at full
scale acceleration.

A voltage to frequency conversion of the accelerometer analog output,
which corresponds to the applied drive voltage on the electrodes, is
well suited to cover the wide dynamic range of the accelerometer 2×10^7
because the measurement bandwidth is limited to a very low frequency.

A first laboratory model has been manufactured and tested during the
last year. In parallel a dedicated pendulum bench has been designed to
decouple the accelerometer during the testing from the seismic
vibrations. The orientations of the bench can be damped or controlled
through different types of servo-controls using magnetic forces and the
informations provided by inclinometers or capacitive sensors.

The Fig. 5 presents a comparison of the outputs of the accelerometer
with a previous model whose solid cubic proof-mass, 3 cm side is made in
silica and weights 70 g.

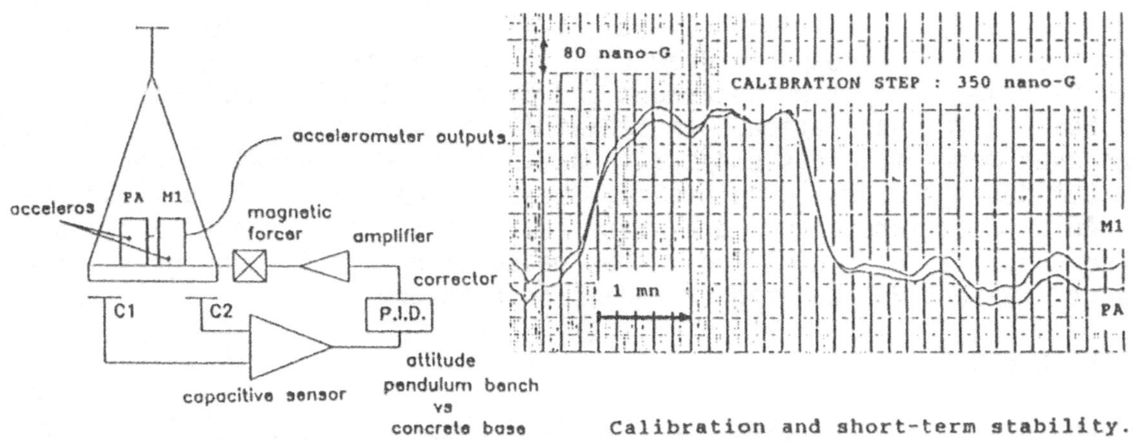

Calibration and short-term stability.

Fig. 5. Pendulum bench inclination controlled through capacitive sensor outputs.

In parallel to these hardware activities, error budgets of the GRADIO accelerometers have been established and lead to the following figures.

- mass   : 5 kg
- volume   : 5 ℓ
- power   : 5 W
- sensitive axes   : full scale range = 10^{-4} ms^{-2}
                                  resolution   for   6 s   integrating   time
                                  = 5×10^{-12} ms^{-2}
- less sensitive axis   : full scale range = 10^{-3} ms^{-2}
                                  resolution   for   6 s   integrating   time
                                  = 5×10^{-10} ms^{-2}
- required thermal stability: 10^{-2} °C.

CONCLUSION

The GRADIO accelerometers that are being developed in ONERA in the frame of the preparation of the ARISTOTELES mission have been optimized for a plane gradiometer accommodated on board a non drag free spacecraft orbiting at an altitude of 200 km.

A resolution of 10^{-2} Eötvös is expected for an integrating time of 6 s. A first laboratory model has been manufactured in 1988 and is being tested in laboratory under normal gravity conditions by means of a dedicated pendulum bench. The development of these accelerometers is now compatible with a first Earth gravity mission at mid-nineties.

Moreover the present status of the GRADIO accelerometer development permits us to envisage other space applications.

In the field of gradiometry, a lunar observer mission for instance can be considered right now. A resolution of a few tenths of one Eötvös appears to be a convenient mission objective because of the lack of knowledge of the Moon Gravity field, because of the presence of mascons and because of the low altitude achievable for the orbiter. Moreover, the absence of atmosphere dramatically reduces the level of disturbances to be rejected by the instrument and by the satellite control.

For Earth observation, the ARISTOTELES mission might open the way to an even more ambitious project involving a drag free satellite.

Besides spaceborne gradiometry, the performances of the ultrasensitive GRADIO accelerometers are very attractive for space applications (Boudon, 1978) involving the measurements of non gravitational accelerations due to drag or radiation pressures.

REFERENCES

Bernard, A. et al. (1988). ARISTOTELES phase A study, ESTEC Contract n° 7528/88/NL/JS Final Report n° 8/3815 PY.

Bernard, A. and Touboul, P. (1987). GRADIO: an electrostatic spaceborne gravity gradiometer. Communication presented at IUGG General Assembly, Vancouver, August 1987, TP ONERA n° 1987-116.

Boudon, Y. et al. (1978). Synthèse des résultats en vol de l'accéléromètre CACTUS pour des accélérations inférieures à 10^{-9} G. Recherche Aérospatiale n° 1978-6.

THE BOUNDARY VALUE PROBLEM APPROACH TO THE DATA REDUCTION
FOR A SPACEBORNE GRADIOMETER MISSION

F. Migliaccio, F. Sansò
Ist. di Topografia, Fotogrammetria e Geofisica - Politecnico di Milano (Italy)
F. Sacerdote
Dip. di Matematica - Università di Pisa (Italy)

1. The target

Our specific target is to achieve the best knowledge of a global gravity field model in terms of coefficients of a truncated expansion in spherical harmonics

$$V \text{ (model)} = \sum_{0^1}^{L} \sum_{0^m}^{1} \sum_{0^\alpha}^{1} v_{lm\alpha} \left(\frac{R}{r}\right)^{1+1} Y_{lm\alpha}(\sigma) \quad , \tag{1.1}$$

$$\left(\; Y_{lm\alpha}(\sigma) = A_{lm} P_{lm}(\phi) \begin{cases} \cos m \lambda & \alpha=0 \\ \sin m \lambda & \alpha=1 \end{cases} \right.$$

A_{lm} normalization constants, $\quad \frac{1}{4\pi} \int Y_{lm\alpha} \, d\sigma = 1 \; ,$

R   mean radius of the earth

L   maximum degree of the expansion   ).

We must underline that what we want to know with the best accuracy are the individual coefficients $v_{lm\alpha}$ and not only their sum.

The state of the art can be illustrated by a plot of relative differences between some global gravity models like Rapp 180 ('86), GEM L2, GEM 10B, GEM T1. The index used is:

$$RD(1) = \sqrt{\frac{\sum_{m,\alpha}(v^1_{lm\alpha} - v^2_{lm\alpha})^2}{\sqrt{\sum_{m,\alpha}(v^1_{lm\alpha})^2 \; \sum_{m,\alpha}(v^2_{lm\alpha})^2}}} \tag{1.2}$$

which is also shown in Fig. 1.1.

To achieve a better knowledge of the $v_{lm\alpha}$ it is possible to use measurements from a spaceborne gradiometer flying, say, at 200 km altitude (almost circular, almost polar orbit), measuring only cross track components to avoid drag problems. We do not want, here, to stress the process by which the observables are formed: so we shall assume directly that V_{zz}, V_{zy}, V_{yy} are measured, say every 5 sec, with z pointing radially and y to east.

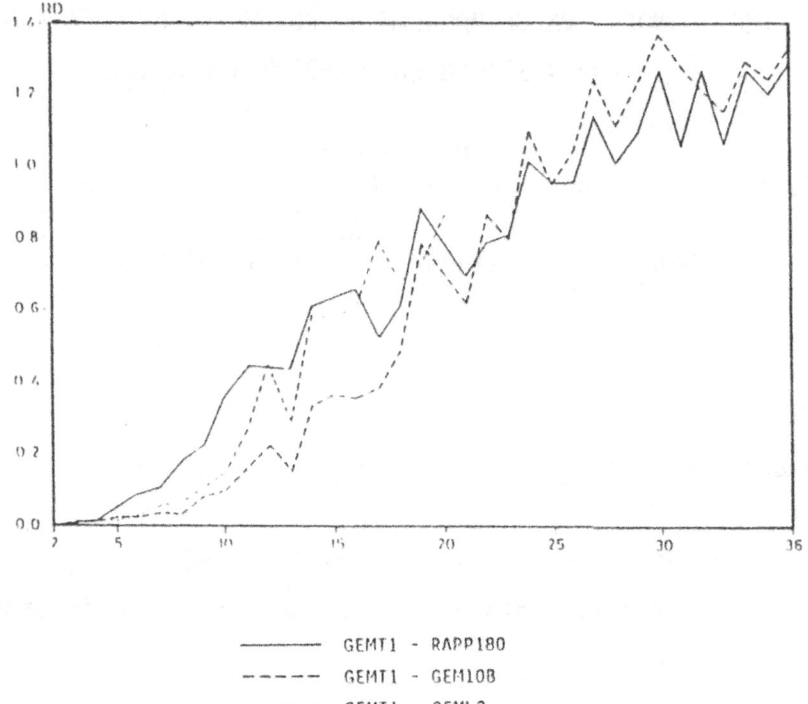

Fig. 1.1

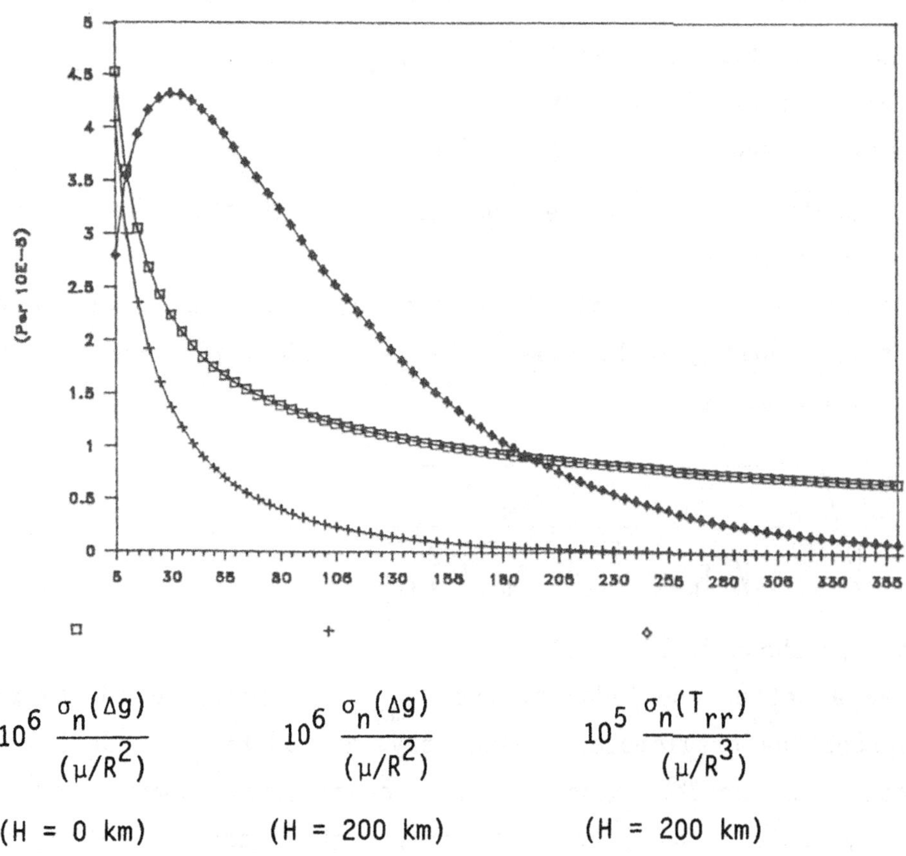

$$10^6 \frac{\sigma_n(\Delta g)}{(\mu/R^2)} \qquad 10^6 \frac{\sigma_n(\Delta g)}{(\mu/R^2)} \qquad 10^5 \frac{\sigma_n(T_{rr})}{(\mu/R^3)}$$

(H = 0 km)   (H = 200 km)   (H = 200 km)

Fig. 1.2

The corresponding observation equations are:

$$\begin{cases} V_{zz} = \dfrac{\partial^2 v}{\partial r^2} + \nu_{zz} \\[2ex] V_{zy} = \dfrac{1}{r \cos \phi} \dfrac{\partial^2 v}{\partial r \, \partial \lambda} - \dfrac{1}{r^2 \cos \phi} \dfrac{\partial v}{\partial \lambda} + \nu_{zy} \\[2ex] V_{yy} = \dfrac{1}{r^2 \cos^2 \phi} \dfrac{\partial^2 v}{\partial \lambda^2} - \dfrac{\mathrm{tg}\,\phi}{r^2} \dfrac{\partial v}{\partial \phi} + \dfrac{1}{r} \dfrac{\partial v}{\partial r} + \nu_{yy} \end{cases} \tag{1.3}$$

with $\nu_{zz}, \nu_{zy}, \nu_{yy}$ temporally uniform and independent noises of 10^{-2} EU per 5 s.
The simplest idea to process these observations in order to get the $v_{lm\alpha}$ coefficients, seems to use (1.1) in (1.3) and apply a least squares estimation approach. However it is known that, due to the non uniform distribution of the observations on the sphere an aliasing happens which conveies power from the coefficients $v_{lm\alpha}$ (l>L), neglected in model (1.1), into the lower frequencies (cfr. [2]). This type of error is not accounted for in usual error analyses, and we do not yet know its amount, which globally should not be very large but it could be strong on specific coefficients.
The target of this paper is to show that it is possible to carry out the estimation of $v_{lm\alpha}$, in an approach of the type used in solving boundary value problems, exploting the observables V_{zz} and V_{zy} and avoiding aliasing.

2. The effects of non uniform distribution

Assume your satellite flies on a perfectly circular orbit with inclination I: accordingly, the longitude of the node Ω has a drift (in an inertial system)

$$\dot{\Omega} \left(= \frac{3}{2} n \frac{J_2}{(1-1^2)^2} \left(\frac{R}{r}\right)^2 \cos I \right)$$

which by the way is almost zero for almost polar orbits.
If $\dot{\theta}$ is the earth's spin, and $n = \mu^{1/2} r^{-3/2}$ is the mean motion of the satel lite, Ω moves along the earth's equator by a quantity

$$\delta\Omega = \frac{\dot{\Omega} - \dot{\theta}}{n} \qquad \text{(in rev)} \qquad . \tag{2.1}$$

Let us assume that this ratio is such that the numbers $k\delta\Omega$ (k=1,2,... Nr of revolutions) are, modulo 1, distributed uniformly on the interval [0,1].
Then, if we approximate the measurement scheme according to the rule: the satellite moves on a circle with fixed inclination I and node at longitude Ω, taking measurements at constant rate; after one revolution the circle changes

suddenly to the next circle with node at longitude $\Omega+\delta\Omega$ and so on.

With this small modification we see that in an area like the one shaded in Fig. 2.1 we expect to have $k=c\delta\Omega$ revolutions and on each of them $n=c\delta\psi$ observations. From the two well-known relations

$$\sin\phi = \sin\psi \sin I$$
$$\cos\psi = \cos\phi \cos(\lambda-\Omega)$$

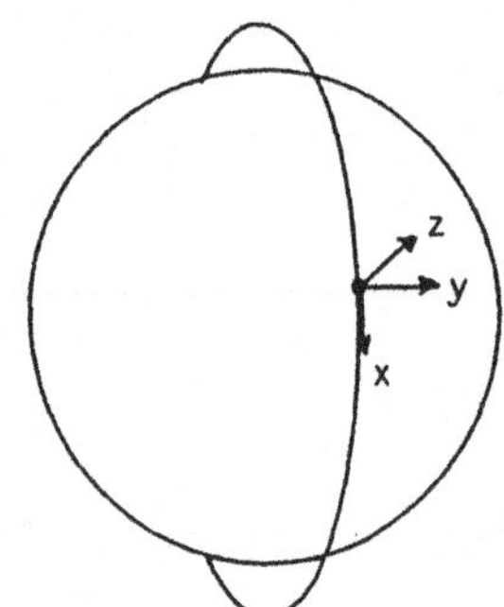

Fig. 2.1

we see that: points with equal ψ have also equal ϕ and moreover

$$d\psi = \frac{\cos\phi}{\cos\psi \sin I}\, d\phi \qquad\qquad d\lambda = d\Omega \qquad \text{(points with equal } \psi\text{)}$$

Using these relations in the expression for the number of observations in dS, we find

$$\delta N = c\, d\psi\, d\Omega = c\, \frac{\cos\phi\, d\phi\, d\lambda}{\cos\psi \sin I} = c\, \frac{dS}{\sqrt{\sin^2 I - \sin^2\phi}} \quad : \tag{2.2}$$

this shows that the observations will tend to distribute on the sphere with density

$$\rho(\phi) = \frac{c}{\sqrt{\sin^2 I - \sin^2\phi}} \quad , \qquad (|\phi| \leq I) \quad ; \tag{2.3}$$

this becomes simply

$$\rho(\phi) = \frac{c}{\cos\phi} \quad , \tag{2.4}$$

for polar orbits. In this situation it has been proved that the l.s. estimates satisfy the "asymptotic" normal system:

$$A\hat{v} = w \tag{2.5}$$

where $\bar{v}$ is the vector of the unknowns $\{\hat{v}_{lm\alpha}\}$. If you take, for instance, V_{zz} as observable, A, the normal matrix with elements $A_{jk\beta,lm\alpha}$, is given by

$$\begin{cases} A_{jk\beta,lm\alpha} = \dfrac{a_j a_l}{4\pi} \displaystyle\int Y_{jk\beta}(P)\, Y_{lm\alpha}(P)\, \rho(P)\, d\sigma \quad , \\[2ex] a_l = \dfrac{(1+1)(1+2)}{R^2} \left(\dfrac{R}{r}\right)^{1+3} \end{cases} \tag{2.6}$$

and w is the known vector with components

$$w_{jk\beta} = \frac{a_j}{4\pi} \int Y_{jk\beta}(P)\, V_{zz}(P)\, \rho(P)\, d\sigma \quad . \tag{2.7}$$

We see that if we believe that the model (1.1) is correct and we insert in it (2.7), we get exactly: $w = Av$, with v the vector of the true coefficients, what together with (2.5) gives $\bar{v} = v$.

On the other hand, the true V_{zz} contains also the degrees $l>L$: if the density $\rho(P)$ was constant, we wouldn't have had any influence from this part because of the orthogonality relations (remember that $j{\leq}L$). Since this is not the case, however, we expect that higher degrees should show up also in $w_{jk\beta}$ $(j{\leq}L)$ and this is precisely the aliasing we want to avoid.

3. The BVP (suboptimal) approach

A different approach to the estimation of $\{v_{lm\alpha}\}$ is the so called BVP approach.

In principle we can schematize the situation by saying that we have several contemporaneous boundary conditions:

$$f_i = B_i v + \nu_i \tag{3.1}$$

affected by independent white noises characterized by a stochastic structure of the type

$$E\{\nu_i(P)\, \nu_j(Q)\} = D(P)\, \delta_{ij}\, \delta(P,Q) \quad . \tag{3.2}$$

The problem (3.1) admits an optimal estimator, i.e. one of minimum variance, according to [5], which however can be difficult to apply in practice.
Nevertheless, it is possible sometimes to perform computations of correct estimates (i.e. not aliased), which, though non optimal, are very convenient from the practical point of view. This happens, for instance, if the boundary operators enjoy a kind of non-mixing property, i.e. calling $X_j \equiv \mathrm{Span}\{Y_{jk\alpha},$ $|k|{\leq}j,\ \alpha=0,1\}$, if it happens that

$$B_i\, X_j \subset g{\cdot}X_j \quad , \qquad \forall\, j \quad , \tag{3.3}$$

with g any given function such that

$$\text{Span}\{g^{-1}X_j, \ j=0,1,2\ldots\} \equiv L^2 \ . \tag{3.4}$$

In this case, in fact, we transform (3.1), for each i, into the equivalent sequence of observation equations

$$\langle g^{-1}Y_{jk\alpha}, \ f_i\rangle = \langle g^{-1}Y_{jk\alpha}, \ B_i v\rangle + \langle g^{-1}Y_{jk\alpha}, \ \nu_i\rangle \ \ ; \tag{3.5}$$

the scalar products being here the simple L^2-products.

Relations (3.3)(3.5) show that we can estimate $\bar{v}_{jk\alpha}$ degree per degree simply by expanding v in series and neglecting the noise term.

In particular, we see that higher degrees in v do not interfere with lower ones, i.e. there is no aliasing in the estimation process.

Proceeding in this way we get a correct estimate $\bar{v}(i) = \{\hat{v}_{jk\alpha}(i); \ j\leq L\}$ for each i up to degree L, and we can afterwards combine these at our best.

To this purpose we just need the covariance matrix of each $\bar{v}(i)$.

$$C(i) = E \{\delta\hat{v}(i) \ \delta\hat{v}(i)^{+}\} \ \ : \tag{3.6}$$

this is derived by covariance propagation from (3.5) and from the covariance of the terms $\langle Y_{jk\alpha}, f_i\rangle$, which, according to (3.2), is given by the formula

$$E\{\langle g^{-1}Y_{jk\alpha}, \ f_i\rangle\langle g^{-1}Y_{lm\beta}, \ f_h\rangle\} =$$

$$= \delta_{ih}\langle g^{-2}(P) \ Y_{jk\alpha}(P), \ D(P) \ Y_{lm\beta}(P)\rangle = \tag{3.7}$$

$$= \delta_{ih} \frac{1}{4\pi} \int g^{-2}(P) \ Y_{jk\alpha}(P) \ Y_{lm\beta}(P) \ D(P) \ d\sigma \ \ .$$

Once $C(i)$ are known, the different estimates $\bar{v}(i)$ can be optimally combined in a weighted mean according to

$$\bar{v} = \{\Sigma_j C_i^{-1}\}^{-1} \ \Sigma_i C_i^{-1}\bar{v}(i) \ \ , \tag{3.7}$$

with covariance $\ C_{\bar{v}\bar{v}} = \{\Sigma_i C_i^{-1}\}^{-1} \ .$

4. An example

We can now show that for the boundary operators (1.3) this approach is good at least for $V_{zz} = B_{zz}v$ and $V_{zy} = B_{zy}v$.

In fact[(*)]

$$
\begin{cases}
B_{zz} Y_{jk\alpha}(P) = a_j Y_{jk\alpha}(P) \\[2ex]
a_j = \dfrac{(j+1)(j+2)}{R^2} \left(\dfrac{R}{r}\right)^{j+3} \quad ,
\end{cases}
\tag{4.1}
$$

and[(**)]

$$
\begin{cases}
B_{zy} Y_{jk\alpha}(P) = \dfrac{1}{\cos\phi}\, b_{jk\alpha} Y_{jk(1-\alpha)} \\[2ex]
b_{jk\alpha} = \dfrac{(-1)^{1-\alpha}\,(j+2)k}{R^2} \left(\dfrac{R}{r}\right)^{j+3} \quad .
\end{cases}
\tag{4.2}
$$

Accordingly we can derive two sets of estimates of $v_{1m\alpha}$, namely:

$$
\bar{v}_{jk\alpha}(1) = \frac{1}{a_j}\, \langle Y_{jk\alpha}, V_{zz}\rangle \qquad (j \leqq L)
\tag{4.3}
$$

and

$$
\bar{v}_{jk\alpha}(2) = \frac{1}{b_{jk\alpha}}\, \langle \cos\phi\, Y_{jk(1-\alpha)}, V_{zy}\rangle \quad , \qquad (j \leqq L) \quad .
\tag{4.4}
$$

In order to derive a unique estimate, we must know the covariances $C(1), C(2)$: to do that we must know what is in our case the density $D(P)$.

To this aim, imagine we have partitioned the whole sphere into small areas dS_i: in dS_i falls a number of measurements

$$
dN_i = \rho(P)\, dS_i
\tag{4.5}
$$

where $\rho(P)$ is given, for polar orbits, by

$$
\rho = \frac{c}{\cos\phi} \quad .
\tag{4.6}
$$

Since dN_i, after integration over the sphere should give the total number of measurements N, we find that

$$
\int \frac{c}{\cos\phi}\, dS = c\, 2\,\pi^2 = N \quad , \qquad \text{i.e.}
$$

$$
c = \frac{N}{2\pi^2} \quad .
\tag{4.7}
$$

(*) Let us observe that we assume to compute the boundary operator on the sphere of radius r.

(**) It is obvious that with V_{zy} we cannot estimate $v_{j0\alpha}$ since $b_{j0\alpha}=0$; the modifications of (3.8)(3.9) are straightforward.

Now we define as "observation" at the point P_i, the centre of dS_i, the average of all the measurements (e.g. V_{zz}) falling in dS_i

$$\bar{V}_{zz}(P_i) = \frac{1}{dN_i} \Sigma \, V_{zz}(P_l) \,, \qquad (P_l \in dS_i) \quad . \tag{4.8}$$

From (4.8) we find for these observables the covariance structure

$$\sigma\{\bar{V}_{zz}(P_i) \, \bar{V}_{zz}(P_j)\} = \delta_{ij} \frac{\sigma_o^2}{dN_i} =$$

$$= \frac{\sigma_o^2 \, \delta_{ij}}{\rho(P_i) \, dS_i} \simeq \frac{\sigma_o^2}{\rho(P_i)} \, \delta(P_i,P_j) \quad . \tag{4.9}$$

Summarizing (4.6)(4.7)(4.9) and comparing with (3.2), we see that:

$$D(P) = \frac{\sigma_o^2}{\rho(P)} = \frac{2\pi^2\sigma_o^2}{N} \, \cos \phi \quad , \tag{4.10}$$

where we remind that σ_o^2 is the variance of each individual observation, and N is the total number of measurements.

With this in mind, and recalling (3.6), we find:

$$C_{jk\alpha,lm\beta}(1) = \frac{2\pi^2\sigma_o^2}{Na_ja_l} \, \frac{1}{(4\pi)^2} \int Y_{jk\alpha}Y_{lk\alpha} \, \cos \phi \, d\sigma \cdot \delta_{km}\delta_{\alpha\beta} \tag{4.11}$$

and also

$$C_{jk\alpha,lm\beta}(2) = \frac{2\pi^2\sigma_o^2}{Nb_{jk\alpha}b_{lk\alpha}} \, \frac{1}{(4\pi)^2} \int Y_{jk\alpha}Y_{lk\alpha}\cos^3\phi \, d\sigma \, \delta_{km}\delta_{\alpha\beta} \quad . \tag{4.12}$$

As we see if we order the coefficients first by $m=0,\ldots,L$, $\alpha=0,1$ and then by $j=m,\ldots,L$, we find that $C(1)$ and $C(2)$ have a block diagonal form, each block being of dimension $L-m$: this makes the matrices $C(1),C(2)$ numerically invertible and formulas (3.7),(3.8) applicable <u>separately</u> for each order m and α.

5. An example of error analysis

Formulas (4.11) and (4.12) can be used to perform the error analysis for simplified models. Since it is not easy to compute explicitly (3.8) with $C(1)$ and $C(2)$, we will make an estimate by using only $C(1)$, i.e. by assuming that V_{zz} only has been used.

If we have estimated a model

$$\hat{v} = \sum_0^L \hat{v}_{lm\alpha} Y_{lm\alpha}(P) \qquad {}^{(*)} ,$$ (5.1)

the so-called commission error is:

$$\delta\hat{v} = \sum_0^L \delta\hat{v}_{lm\alpha} Y_{lm\alpha}(P) \quad .$$ (5.2)

If we want to use a smoothed model, i.e. if we apply a moving average over a disk of angular radius ψ, we can obtain it simply by multiplying the coefficients by the β_l factors, i.e.

$$\bar{v} = \sum_0^L \beta_l \hat{v}_{lm\alpha} Y_{lm\alpha}(P) \; , \quad (\beta_l = \frac{1}{2l+1} \frac{1}{1-\cos\psi} \{P_{l-1}(\cos\psi) - P_{l+1}(\cos\psi)\}) : $$ (5.3)

to this corresponds a pointwise commission error $\quad \delta\bar{v} = \sum_0^L \beta_l \delta\hat{v}_{lm\alpha} Y_{lm\alpha}(P).$

We define now a mean square commission error as

$$E\{\frac{1}{4\pi} \int \delta\bar{v}^2 d\sigma\} = E\{\sum \beta_l^2 \, \delta\hat{v}_{lm\alpha}^2\} =$$

$$= \sum_0^L \beta_l^2 \, \Sigma_{m,\alpha} \, C_{lm\alpha,lm\alpha} =$$

$$= \frac{\sigma_o^2}{8N} \sum_0^L \beta_l^2 \frac{1}{a_l^2} \, \Sigma_{m,\alpha} \int Y_{lm\alpha}^2 \cos\phi \, d\sigma =$$ (5.4)

$$= \frac{\sigma_o^2}{8N} \pi^2 \sum_0^L \beta_l^2 \frac{(2l+1)}{a_l^2}$$

Just for the readability of the result, one can divide $\bar{v}$ by γ, getting in this way an averaged model for the geoid.

Correspondingly, one can define a mean square error per degree

$$\eta_l^2 = \frac{\sigma_o^2 \, \pi^2}{\gamma^2 \, 8N} \, \beta_l^2 \, \frac{(2l+1)}{a_l^2}$$ (5.5)

and the global mean square commission error

$$\mathcal{E}_L^2 = \sum_0^L \eta_l^2 \quad .$$ (5.6)

(*) Observe that with (5.1) $\hat{v}$ is computed at ground level, i.e. for r=R.

The plots of η_1 and $\mathcal{E}_L$ up to degree 360 in units of cm, can be found in Fig. 5.1, Fig. 5.2: the errors have been computed putting $\sigma_0 = 10^{-2}$ E.U. per 5 sec, $N = 3,110,400$ corresponding to 6 months of mission, ψ corresponding to a $1^0 \times 1^0$ block at the equator. With some rough reasoning, it is also possible to guess that the so-called omission error, namely:

$$\left\{ \sum_{L+1}^{+\infty} \frac{\beta_1^2}{\gamma^2} \Sigma_{m,\alpha} v_{1m\alpha}^2 \right\}^{1/2} \qquad \text{should be in the order of few centimeters.}$$

Commission error per degree

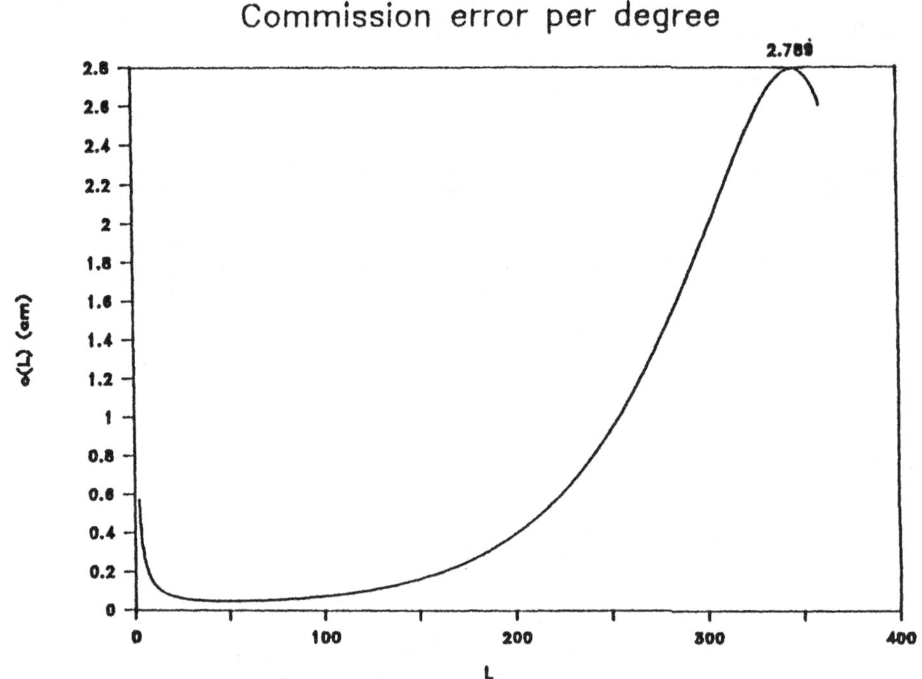

Fig. 5.1

Cumulated commission error

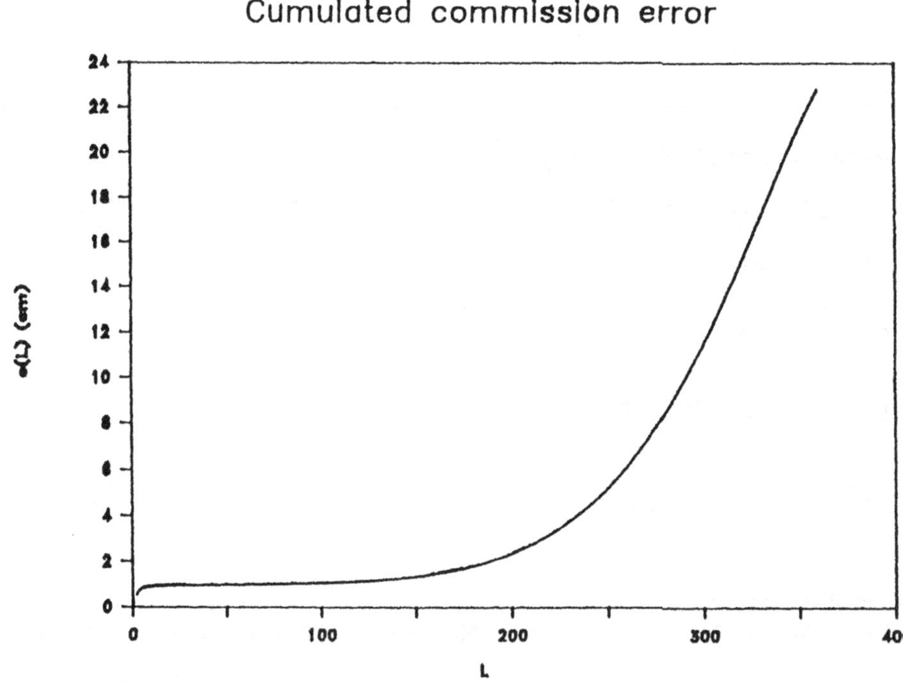

Fig. 5.2

6. Comments

- It is important to stress that the estimates devised by (4.3),(4.4) are both
unbiased and correct and free of aliasing: however we have not fully exploited
our information since V_{yy} doesn't seem to fall in the group of functionals for
which it is easy to find the orthogonal sequence for $\{Y_{lm\alpha}\}_{yy}$, as it has been
for V_{zz}, V_{zy} (cfr. (3.5)).

- The numerical results of the error analysis are similar to those reported in
literature. Moreover, one should be aware that there will be many gaps in the
data due to periods of manovering, a complete lack of data at the poles due to
a non perfect polar orbit, and, mostly important, for instrumental reasons
there will be a very high noise at low frequencies (below 5 mHz).

- It is maybe possible to recover a good knowledge of the low frequency part
of the signal, from phase data of a GPS in the same pay-load with the
gradiometer. However, also for phase data the aliasing problem maybe harming.
Whether it could be possible to treat these data as well in a BVP approach, is
an interesting question.

References

[1] O. Colombo (1989): The Global Mapping of the Gravity Field with an Orbit
Full Tensor Gradiometer: an Error Analysis. Proc. XIX IUGG Gen. Assembly,
Vancouver, Canada.

[2] F. Migliaccio, F. Sansò (in print): Data Processing for the Aristoteles
Mission.

[3] R. Rummel, O. Colombo (1985): Gravity Field Determination from Satellite
Gradiometry. Bull. Gèod., vol. 59.

[4] F. Sacerdote, F. Sansò (1985): The Overdetermined Boundary Value Problems
of Physical Geodesy. Manuscripta Geodaetica, vol. 10, n. 3.

[5] F. Sansò (1987): The Wiener Integral and the Overdetermined Boundary
Value Problems of Physical Geodesy. Manuscripta Geodaetica, vol. 12, n. 4

[6] F. Sansò (1989): On the Aliasing Problem with the Harmonic Analysis on
the Sphere. II Hotine-Marussi Symposium, Pisa, Italy, 3-5 June.

APPROACH ON SATELLITE GRAVITY GRADIOMETRY AND ITS VISTAS OF APPLICATIONS

Meng Jiachun and Cai Ximei
Institute of Geodesy and Geophysics
Chinese Academy of Sciences, 54 Xu Dong Road, Wuchang, Hubei, China

ABSTRACT

Satellite gravity gradiometry and its instrumentation is an ultra-sensitive detection technique of space gravity gradient being developed internationally now. Gravity gradiometry will be of great use for detecting the short wave component of the earth's gravity gradient field. In this paper the necessity and possibility of carrying out the satellite gravity gradiometry and developing the ultra-sensitive gradiometer are discussed with regard to the theoretical characteristics of gravity gradiometry, very high sensitivity of tensor gradiometer, accurate geophysical interpretation and the vast vistas for applications. This paper demonstrates systematically the problems of refinement of the earth's gravity field model, and deals with the great significance of gravity gradiometry in inertial navigation, gravity survey, geodynamics and earthquake prediction research. In the meantime, it expounds its importance in monitoring the microgravity environment, examining the gravity field continuation theory, improving the analyses subjected orbital vehicle to gravitational force and exactly testing the law of gravity.

I. PUTTING FORWARD THE QUESTIONS

Satellite gravity gradiometry is a new domain of studying the characteristics, fine structure and variation process of the earth's gravity gradient field as well as its sensing technology. Installing a tensor gradiometer on satellite or in space station to determine the gravity gradients can reveal the information and influence of the earth's gravity field. Thus satellite gradiometry is a science of studying the theory, methods and instruments of gravity gradiometry in space and its applications, and it is a new subject being developed rapidly with the beginning in space times. At present it is still at the stage which carries out actively the developing of the ultra-sensitive gradiometer and pre-research of the applications.

Eötvös' torsion balance is a earlier period instrument of the gravity gradiometer. It had been used to determine the spatial variations of gravity by Eötvös, especially the curvatures of the equipotential surface of gravity. In 1920's torsion balance had widely been used for geophysical exploration. Later on it has been substituted by sensitive and more economical gravimeter. Until the beginning of space times the designers of spacecraft discovered that the use of gravity gradiometry for space navigation will be more suitable in free fall orbit. From then on, the developments of gravity sensor and gradiometer of various principles have made rapid progress. The accuracy of the assembly of modern inertial navigation has come up to such standards, i. e. the limits of improving the accuracy further doesn't come from the performance of gyroscope and accelerometer and from vertical deflection. Improving the hitting accuracy of weapon system is more efficacious than increasing the equivalent of warhead. Moreover the hitting accuracy of long-range weapon will depend finally on a understanding degree of the earth's gravity field, i. e. on the determinable accuracy of geoid undulation and vertical deflection. In addition, the objectives of geodynamics project are as follows: (1) to contribute to the understanding of the solid earth, in particular the processes that result in movement and deformation of tectonic plates; and (2) to improve measurements of the earth's rotational dynamics and its gravity and magnetic fields. Just due to the requirement made by the development of society, science and technology, this impels us actively to carry out the projects of Geopotential Research Mission (GRM).

In order to achieve the GRM project, various theory and instruments being used to determine the global gravity field are being considered over the years. In the earth's orbit the tensor gradiometer is able to sense and record the spatial second derivatives of gravitational potential. Satellite gravity gradiometry (SGG) is the most promising means of mapping the short-wave information of the earth's gravity field. Moreover, SGG have the characteristics of the strict theory, extra-high sensitivity, accurate explanation and wide uses etc. Consequently it will become the most advanced techniques in the development of dynamical gravimetry in the coming tens years. Therefore since late in 1960's, gravity gradiometry has been taken seriously by many countries.

II. CHARACTERISTICS OF SATELLITE GRAVITY GRADIOMETRY

Now according to the theoretical characteristics, instrument performance and observation environment of the gravity gradiometry the advantages of satellite gradiometry have been comprehensively investigated as follows:

1. Not affected by inertial acceleration. For gravity vector the gravitational force and inertial centrifugal force can been separated by using only gravity gradient tensor.

Thus in marine and aerial gravimetry, by a combination of gravimetry with gradiometry, it is theoretically rigorously possible to eliminate the inertial noise (Moritz, 1985).

2. To determine real time vertical deflection and improve the accuracy of navigation. In modern times, the uncertainty of gravity has become a important source of error of high quality inertial navigation system which used for modern spacecraft and submarine. A review of inertial system survey performance indicates that addition of a gradiometric sensor could significantly improve vertical deflection survey accuracy. To accomplish a similar reduction in gravity vector survey error without gradiometers, it would require more than a tenfold increase in gyroscope and accelerometer performance (Heller, 1981).

3. There is basically no Eötvös effect.

4. To be insensitive to altitude change. Gravity gradient isn't fundamentally affected by the vertical acceleration of the spacecraft.

5. Extra-high sensitivity of the instrument. At present, generally speaking, the sensitivity goals for conventional gradiometer are $10^{-2}E$, and the sensitivity goals for cryogenic gradiometer will be $10^{-4}E$.

6. Improvements of null stability and scale factor stability. The principal source of error of the room-temperature instruments is mechanical hot random noise. And the use of liquid helium itself provides a stable, gradient-free, thermal environment. Thus , it is estimated that in a superconducting instrument the secular drift will be small , scale factor stability be better, and electromagnetic shielding be good (Paik, 1981).

7. Disturbance in space is smaller than on the terrestrial. Since any gravity sensor is inevitably an accelerometer, vibration isolation is a serious problem in an earth-based instrument. In an earth orbit, the sensor is essentially decoupled from external vibrations and the first derivative of the gravitational potential is balanced by the geodesic motion of the satellite to one part in 10^8, a limit due to radiation pressure and aerodynamic drag on the satellite.

8. Mapping rapidly the global gravity field. Robbins (1985) deems that the mean accuracy of $30' \times 30'$ is about 6 mgal when the duration of flight is 20 days, and it will decrease to around 5 mgal while the duration nears 6 months.

9. Accurate geophysical explanation. Geophysical explanation made by gravity tensor or gravity vector will be more accurate than by gravity scalar quantity.

10. Model-independent recovery of the full gravity field. A tensor gravity gradiometer can give simultaneously a three-axis measurement information, Moreover, the redundancy provided by the three-axis measurement allows noise reduction and model-independent recovery of the full gravity field (Paik, 1981).

III. THE VISTAS OF APPLICATIONS OF SATELLITE GRADIOMETRY

Now we discuss the wide uses of satellite gradiometry with respect to inertial navigation, the model of the earth's gravity field, geophysical exploration, geodynamics and earthquake prediction research as well as space sciences.

1. Inertial Navigation

Many researchers indicate that the "directions" in space will be shown by gravity gradiometer. Space navigation can be carried out by using the gravity gradiometry.

In respect of gravity vector, to take advantage of Riemann curvature tensor is only method which can keep gravitational force separate from inertial force. Therefore using the principle involving gradiometry, we can get a purely inertial signal, consequently we can give theoretically rigorous position differences (Moritz, 1985).

The tensor gravity gradiometer constitutes an integrated package of inertial navigation instruments. The local vertical could be determined continuously using the tensor gradiometer outputs, the vehicle velocity and position in three dimensional space could be obtained integrating the linear acceleration signals, and the angular orientation of the vehicle at each moment may be computed from the outputs of the angular accelerometers (paik, 1981). For this reason, the tensor gradiometer should be capable of monitoring attitude and position directly.

2. The Model of the Earth's Gravity Field

The model of the earth's gravity field is one of the important problems of space environment model research, it is also the important contents of studying fine structure of the earth's gravity field. Usually the model of the earth's gravity field is to be expressed in terms of spherical-harmonic expansion. So far, the gravity field models recommended have been multitudinous, among them GEM 10B, GEM 10C, OSU81, OSU86C/D etc are good results. Both models GEM 10C and OSU81 have the potential coefficients completed to degree and order 180. It is generally considered that these two models could reflect the local gravity field.

However, the existing gravity field models of the earth remain still to be refined. Ning Jinsheng et al.(1985) have shown clearly that though the models of OSU81 and GEM10C have high degree coefficients, they could not well reflect the situation of the local gravity field in China, especially they can not improve the accuracy because of considering the high degree terms. Therefore we must combine it with surveyed data in our country to improve the geopotential coefficients.

Bhattacharji (1984) held that the improvements of topographic effect for existing earth's gravity field models (GEM10B/C, OSU81) would be needed further to make the precise determination of global $1^\circ \times 1^\circ$ mean gravity anomalies and geoid undulations. In addition, ground techniques and conventional satellite methods cannot be used much further to improve the global accuracy and resolution of the earth's gravity field models.

If the earth's gravity field models won't be improved significantly, then the studies concerned with it aren't able to bring all their potential into full play. For example, at present one of the main error sources of determining ERP (Earth Rotation Parameter) by SLR (Satellite Laser Ranging) technique is the mechanical model. Particularly the imperfect earth's gravity field model. Calculating the accuracy of the stress field of mantle convection currents under lithosphere depends on the refinement of the earth's gravity field model. Therefore, some famous experts in astrometry, geodesy, geophysics, and inertial navigation expect that the existing earth's gravity field models will be further improved.

The gravity gradiometry is rather sensitive to short wave components of larger density difference. If the sensitivity of gradiometer is accurate to 0.01E, then the spherical harmonic potential coefficients reach degree and order 75 in resolution in the integrated time of 30 sec. When the orbit altitude is 200km, if the accuracy of gradiometer is 10^{-3} E, the potential coefficients can be estimated at degree and order 230; if it is 10^{-4} E, then the coefficients can be estimated to be degree and order 300 (Colombo et al., 1983).

In addition, Glaser and Sherry (1972) demonstrated that every method (Doppler, altimeter and gradiometer) had its applicable range. Thus we should utilize synthetically all kinds of gravity detection to ascertain the model of the earth's gravity field. The spherical-harmonic coefficients up to degree 35 could been obtained by Doppler tracking method, and the coefficients over degree 35 could been calculated by gravity gradient method.

3. Geophysical Exploration

In solid geophysics the domain needed to increase and improve the gravity information and the geopotential knowledge is numerous.

It is well known that the gravity observations in the earth's surface are some values of discrete points. To determine the geoid undulation, the perturbation of satellite orbit, and the vertical deflection of the fall point etc, it often needs to know the mean gravity anomalies of the blocks (for example $5' \times 5'$, $1^\circ \times 1^\circ$ etc) in many local region and the whole world. It is much to be regretted that the collocation of gravity anomalies would hardly be made up by computational method for the region of the rugged topography and the one that observation point distribution is gathered in a corner of the block. For default of observed gravity data or entirely blank areas, in the medium and high mountain region of varied topography in our country, the maximum difference between the mean free-air gravity anomalies calculated by the isostatic theory and the observed gravity data in $1^{\circ} \times 1^{\circ}$ blocks is about 7 mgal, the standard deviation is ± 5.22 mgal (Meng et al., 1981). Wichiencharoen (1985) warned that 2.5 mgal may be obtained for an area with a smooth gravity field, while only 30 mgal may be obtained, if the gravity field is rough.

Now satellite gravity gradiometry needs to explore gravity variations in the resolution range of $50-200$ km, over lands as well as over oceanic parts. But for the resolution power of 50km the sensitivity of gradiometer will be $10^{-4} \, \mathrm{EHz}^{-1/2}$.

Robbins (1985) has obtained that satellite gradiometry data at an altitude of 200km can be used successfully for the determination of $30' \times 30'$ mean gravity anomalies to an accuracy of 9.4 mgal from his algorithm. Global studies by Rapp (1987) indicate an about 120 percent decrease in the sensitivity with respect to $1^{\circ} \times 1^{\circ}$ mean geoid undulations when changing altitude from 160km to 200km, corresponding to an increase in the $1^{\circ} \times 1^{\circ}$ mean geoid error from 3.4cm to 7.8cm. In absolute numbers, the increase is from below 5cm to nearly 10cm. This shows that a 160km altitude is essential for obtaining the oceanographic applications to need accuracy of 5cm. Otherwise important oceanographic goals can not be reached.

4. Geodynamics and Earthquake Prediction Research

According to four main missions of geodynamics and earthquake prediction research, it is necessary to establish:
  - global and continental network of reference points;
  - regional network of high earthquake activity area,
and
  - to improve the knowledge of the earth's gravity field in order to identify its characteristics of short wave (probably is $100-150$km).

In accordance with the above basic purposes and requirements, to put it briefly, it would want to give the precise position of terrestrial points and the characteristics of short wave of gravitation. Satellite gravity gradiometry may satisfy the demands in following aspects because position accuracy will be of extra high and be the most effective on shorter wave.

1)To Support the Terrestrial Geodesy. Seeing that the capacity to determine the geocentric distance of a spacecraft in the vicinity of the earth is better than a decimeter, and that a gradiometer also provides the means for obtaining radial distances in a way that is quite independent from tracking (and tracking station) errors, thus it may been expected that satellite gradiometry can contributed to earthquake prediction research in the following way:

• Independent check on conventional methods

Satellite gradiometry could be used to check the geocentric position of terrestrial stations obtained by other means, and to improve the coordinates of tracking stations and thus the effectiveness of satellite-positioning systems that depend on those coordinates, such as GPS, etc.

• Setting up three-dimensional global network

Combined with other accurate positioning techniques that can determine horizontal distances, such as VLBI, satellite gradiometry could be used to set up a three-dimensional global network where distances are known in all directions to better than one decimeter.

2) *Improving the Determination of Satellite Orbit*. Now the accuracy of the altimeter approaches one decimeter rms. Improvement of the accuracy of the altimeter itself is of limited use if the radial orbital errors cannot be brought well below one meter. In the future, it may be considering to launch a spacecraft carrying a gradiometer together with the altimeter. In addition to its capacity for detecting radial errors, the accelerometers of the gradiometer can sense the forces of drag and solar pressure, thus improving the calculated orbits beyond what can be obtained by conventional tracking alone (Colombo et al. , 1983).

3) *Probing into the Stress Field of Mantle Convection Currents under Lithosphere*. Many studies concerning the stress field of mantle convection currents have been made by some learned men in our country in terms of Runcorn's(1964) method. For example, the stress field map of low-degree and high-degree of mantle convection currents under lithosphere of the South China Sea has strongly shown the formation of the South China Sea waters and its geological structure zones understanding now . The stress field of mantle convection currents in the Xizang Region has indicated that there is a up welling current as a centre in the middle part of the Xizang Region , and it is closely related to geography of the earth's crust surface .

4) *The Long-Time Change of Gravity Anomaly*. Satellite gravity gradiometry is a much efficient measures to determine the short-wave informations of the gravity field directly, and it has simultaneously also a capabilities to solve a space-time problem of gravitational force.

5. Space Sciences

Applications of gravitation in space sciences can make an investigation on detecting the micro-gravity environment, examining the continuation theory of gravity field, improving the analyses subjected vehicle orbit to gravitational force, and testing exactly the law of gravity etc.

1) *Detection of the Micro-Gravity Environment*. Micro-gravity environment is the most main aspect in space natural resources. The uses of space natural resources leading by micro-gravity have set up the researches on space micro-gravity sciences now. For example, space life sciences, pharmaceutical processing in space, and materials processing in space are mainly to be made in use of the micro-gravity environment. Up to now, the quantitative descriptions that the state of organism is affected by different gravity and micro-gravity environment are wanting investigation yet. In pre-research work for space material sciences it is also necessary " to understand the influence of different levels of micro-gravity on various material test and the range being able to test". The overall tentative ideas of applications of space station in our country have put forward that "making demands on monitoring of the environment between the sun and the earth will be raised to the quantitative descriptions; real time prediction service and mechanism research of mutual effect process between the sun and the earth will be engaged in". It is obvious that satellite gradiometry could satisfied the above-mentioned requirements.

2) *Examining the Continuation Theory of Gravity Field*. The continuation of gravity field outside the earth means mainly the mathematical method problems in which gravitational field in the earth's outerspace could be calculated from terrestrial observation data. It is usually referred to as the upward continuation problems too. The upward continuation problems have various solutions. The gravity disturbance vector in space obtained by various methods generally could only determine which is better by using the comparative analysis methods (Meng et al., 1986). However, satellite gradiometry could directly determine the horizontal and vertical component of gravitation at any point in space with an ultra-sensitive gravity gradiometer. This would provide a possibility testing the continuation theory of the earth's gravity field.

3) *Improving the Analyses of Subjecting Vehicle Orbit to Forces*. The disturbing gravitational force in space is generally computed from terrestrial observation data or the gravity field model. Because gravity gradiometry is more sensitive to the gravity disturbance field with larger density difference, and it could more conform to reality than the values obtained by using discrete values of gravity points at the earth's surface and isostatic theory, therefore, the use of satellite gravity gradiometry could efficiently improve the analyses of subjecting vehicle orbit to forces.

4) *Null Test of the Gravitational Inverse-Square Law*. H. A. Chan et al.(1982) have verified this condition within the following experimental uncertainty.

$$\sum_i \Delta\Gamma_{ii} = (+0.15 \pm 0.23) \times 10^{-9} \text{sec}^{-2}$$

REFERENCES

Bhattacharji, J. C. (1984).Improvements on existing geopotential coefficients models for precise determination of 1° global geoid and mean gravity anomalies, *Bull. Géod.* 58, 31−36.

Colombo, O. L. and Kleusberg, A. (1983). Applications of an orbiting gravity gradiometer, *Bull. Géod.* 57, 83−101.

Heller, W. G. (1981). Prospects for gradiometric aiding of inertial survey systems, *Bull. Géod.* 55, 354−369.

Meng Jiachun et al., (1981). Calculation of mean free-air gravity anomalies of 1° × 1° squares by applying the isostatic theory, *Acta Geodaetica et Geophysica*, No. 3, 97−107.

Meng Jiachun et al.,(1986). Comparison of the methods for determining the gravity disturbances outside the earth, *Acta Geodaetica et Geophysica*, No. 7, 84−102.

Moritz, H. (1985). Inertia and gravitation in geodesy, Proceedings of the Third International Symposium on Inertial Technology for Surveying and Geodesy, Banff, Canada, September, 16−20, 1985, K. P. Schwarz (ed.).

Ning Jinsheng and Li Kun, (1985). The computation of high degree spherical harmonic expansions of geopotential and their adaptability in China, *IGG Special Publication* No. 8, 51−59, Institute of Geodesy and Geophysics, Chinese Academy of Sciences.

Paik, H. J. (1981). Superconducting tensor gravity gradiometer for satellite geodesy and inertial navigation, *Journal of Astronautical Sciences*, 29, 1, 1−18.

Robbins, J. W. (1985). Least squares collocation applied to local gravimetric solutions from satellite gravity gradiometry data, *Dept. of Geod. Sci. Rep.* No.368, Ohio State Univ.,Columbus.

Contribution of the Gravity Probe B Mission
to Geodesy and to Satellite Navigation

Mark B. Tapley
Graduate student, Aeronautics and Astronautics
Stanford University, Stanford, CA. 94305 USA

John V. Breakwell
Professor Emeritus, Aeronautics and Astronautics
Stanford University, Stanford, CA. 94305 USA

C. W. F. Everitt
Professor in the High Energy Physics Laboratory (Research)
Principal Investigator, Gravity Probe B, Stanford

The Gravity Probe B (GP-B) experiment, under development at Stanford University for over twenty years, is designed to measure with high precision two extremely important effects of Einstein's general theory of relativity as observed with gyroscopes in Earth orbit. These are respectively the frame-dragging effect exerted by a spinning body (the Earth) on an angular momentum vector (the set of gyroscopes), and the relativistic "geodetic" effect resulting from the motion of the gyroscope in its orbit through the curvature in space-time produced by the Earth's mass. Because the frame-dragging effect is very small - in this case, only 0.042 arcseconds per year as defined with respect to the line of sight to a distant guide star - we must meet several stringent conditions in order to measure it to the 1% precision or better that is the goal of the experiment. In particular we need make the Gravity Probe B spacecraft "drag free" and need carry on it a GPS (Global Positioning System) receiver for precise orbit tracking. In doing this in a satellite of relatively low altitude (650 km) in polar orbit around the Earth, we create a system that will also, as an unexpected "coexperiment", improve knowledge of the Earth's gravity field by a factor of 100 to 1000.

The requirement that the satellite be "drag free" stems from the smallness of the relativistic effect. The largest non-relativistic perturbations on the gyroscopes are from suspension forces; to minimize these, we need place the gyroscopes in a very low-acceleration environment. Therefore, the satellite carries a control system which compensates for the effects of atmospheric drag, radiation pressure, and other surface forces, down to a level of 10^{-10} g.

The polar orbit is dictated by consideration of the relative magnitudes of the two relativistic effects. Since the "geodetic" effect is about 150 times larger than the frame-dragging effect, we need a reliable way to separate them. Because one acts in the plane of the orbit and the other perpendicular to this plane only for a polar orbit, we choose such an orbit.

The mission design features each contribute to the ability of the satellite to measure the Earth's gravity field. The drag-free control system, by compensating for non-gravitational perturbations, allows long-arc solutions and minimizes non-gravitational aliasing. In addition, telemetry of the control signals allows accurate measurement of nongravitational forces, contributing to aeronomy and other studies. The polar orbit insures that the satellite moves over every portion of the Earth so that its orbit is affected by deviations in the gravity field everywhere, even over the poles, enabling us to solve for the complete gravity field. The GPS system allows accurate, continuous and evenly distributed measurements over the entire surface of the Earth. This is an improvement over previous geodesy satellites for which measurements could only be taken when the satellite was in view of a ground station. Finally, the satellite's relatively low altitude (it has a semi-major axis smaller than any of the satellites used in GEM-T1) will maximize its sensitivity, especially to higher-degree coefficients in the gravity field.

We attempt here to quantify just how well the GPB satellite will measure the gravity field. The analysis we apply is one by Breakwell (Breakwell, '79). This is the fairly well-known "flat Earth" treatment.

The basic development of this treatment is as follows: We use a convolution in two dimensions to approximate the Poisson equation extending the potential field of the Earth above the surface. The approximation applies only if h, the altitude of the satellite, is much less than the radius of the Earth. If we use a Fourier transform in two dimensions (x and y) of the potential, where

$$U(\vec{\omega},h) = \int\int_{-\infty}^{\infty} e^{-j(\omega_x x + \omega_y y)} \, U(x,y,h) \, dx \, dy$$

then Poisson's equation reduces further to

$$U(\vec{\omega},h) = e^{-\omega h} \, U_0(\vec{\omega})$$

$$\text{where} \quad \omega = \sqrt{\omega_x^2 + \omega_y^2} = |\vec{\omega}|$$

$$U_0(\vec{\omega}) = U(\vec{\omega},0)$$

This simplifies the error analysis considerably. We assume some a priori information on the gravity field. Breakwell showed that if we take a measurement vector ξ, related to the potential by a multiplier vector H,

$$\xi(\vec{\omega}) = H(\vec{\omega}) \, U(\vec{\omega}) + W$$

where W represents white noise, then a minimum variance analysis gives the average reduction in the uncertainty in the coefficients of U at degree *l* as

$$F = \sqrt{\frac{1}{2\pi} \int_0^{2\pi} \frac{d\alpha}{1 + \phi_{U_0}(\vec{\omega}) \, H^T(-\vec{\omega}) \, \Phi_W^{-1} H(\vec{\omega})}}$$

where α is the direction of the vector
$$\vec{\omega} = (\omega_x, \omega_y)$$

$\phi_{U_0}(\vec{\omega})$ is the *a priori* knowledge spectrum

Φ_W is a diagonal noise level matrix

ω **is a spatial frequency given by the degree l divided by the Earth's Radius**

We assume the noise level matrix Φ_W is an identity matrix multiplied by

$$\sigma_\xi^2 * 2\pi^2 R^2 / N$$

where N is the number of vector measurements, σ_ξ^2 is the measurement variance of each component, and R is the radius of the Earth.

$\phi_{U_0}(\omega)$, represents the *a priori* knowledge of the field as a function of ω only. It is taken to be $4\pi\mu_\oplus^2\sigma_l^2$ where σ_l^2 is the average variance of the GEM-T1 model at degree l.

The vector H depends on the type of measurement. For position measurements (north-south, east-west, and vertical), H is given by

$$H(\vec{\omega}) = e^{-\omega h} \left(\frac{1}{j\,V^2\omega_x}, \;\; \frac{\omega_y}{j\,V^2\omega_x^2}, \;\; \frac{\omega}{V^2\omega_x^2} \right)$$

Breakwell showed that, for velocity measurements, three simultaneously visible satellites in random directions could provide the same information as x, y and h measurements of velocity. The same principle applies to position measurements; twelve GPS satellites would be visible at any time, and the information they provide would be equivalent to four position measurements each, in the x, y, and h directions. We assume one measurement each second from each visible GPS satellite, over a two-year mission, so that N = 63,115,200. Finally, we assume the accuracy of the measurements (σ_ξ) to be approximately 2 cm. We believe the 2-cm value to be pessimistic in specifying the precision of position measurements; actually precisions of 0.3 cm are likely (Eschenbach). However, for all position measurements in any arc, a phase ambiguity exists which could cause the position to be in error by an integer number of 10-cm wavelengths. The phase ambiguity can almost certainly be resolved due to the length of the measurement arcs and the geometry involved (Tapley et.al.). If this is not so, the result would be a loss in information at low frequencies (degree less than about 12) only.

We show the result of calculations using these assumptions in Figure 1. Here the vertical axis is the ratio of uncertainty (variance) in coefficients of that degree to the corresponding uncertainty in GEM-T1. The horizontal axis is degree l. The graph shows that the expected improvement is a factor of 1000 over the range l = 5 to l = 15, and a factor of 100 up to l = 30.

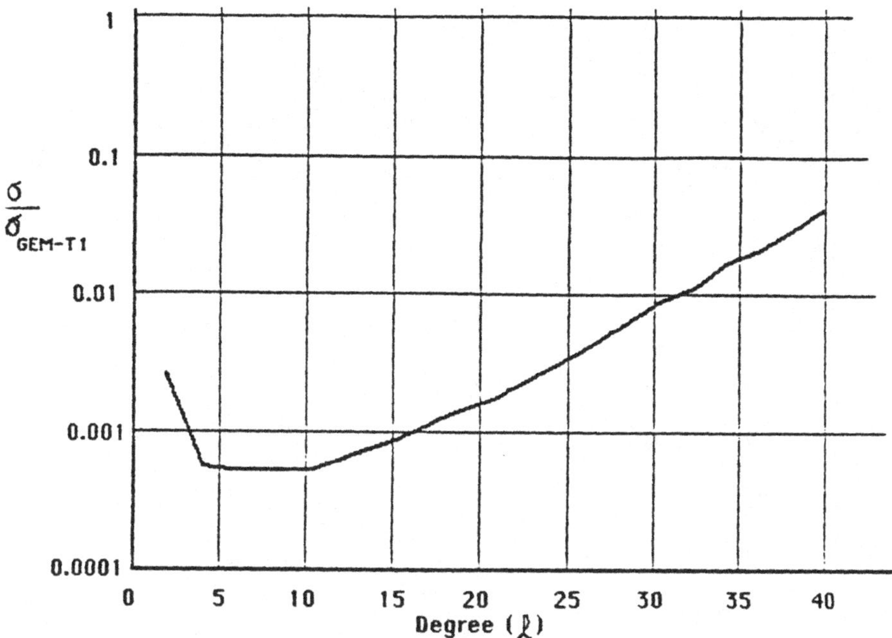

Fig. 1. Average Reduction factor in variance of Geopotential coefficients as a function of degree.

The size of this improvement is somewhat unexpected, but becomes less surprising if one examines the applicable features of the mission. The low altitude of the orbit and the drag-free control system contribute strongly, but the unusually large amount and high quality of the measurements is even more important.

Figure 2 shows these results in terms of centimeters of geoid error variance. Here the "Satellite Only" curve represents GEM-T1, and the "Satellite + Altimetry" curve represents the predicted improvement when radar altimetry data from TOPEX is combined with the satellite data used in GEM-T1. The expected performance of the GP-B mission is shown by the three lines labeled GP-B. The three lines represent orbit altitudes of 650, 700, and 750 km.

The curves labeled "Ocean Topography Signal" (Nerem, Tapley, and Shum), "M2 Tide" (Williamson and Marsh), and "Glacial Rebound" (Peltier) show the relative magnitudes of some geophysical phenomena of interest. All of these contain time-dependent elements, which may make resolving them more difficult than the graph would indicate. Nevertheless, the planned TOPEX mission, for example, would clearly benefit greatly from the increase in ability to distinguish ocean topography signals from the geoid.

Figure 3 shows more clearly the effect of altitude changes. The curves represent contours of constant information; anywhere along one of these

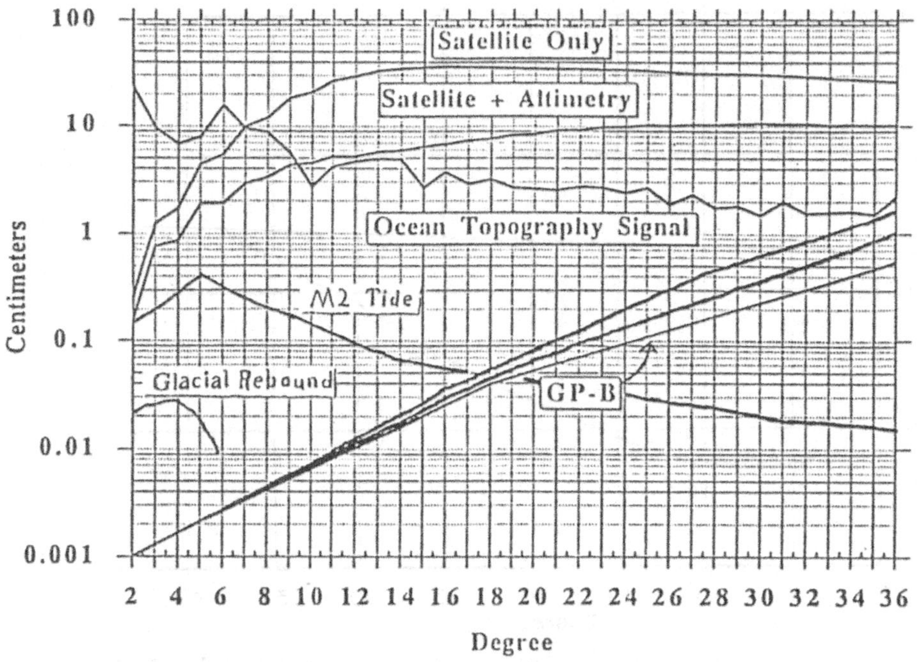

Fig. 2. Geoid Error Variace and Geophysical Signals

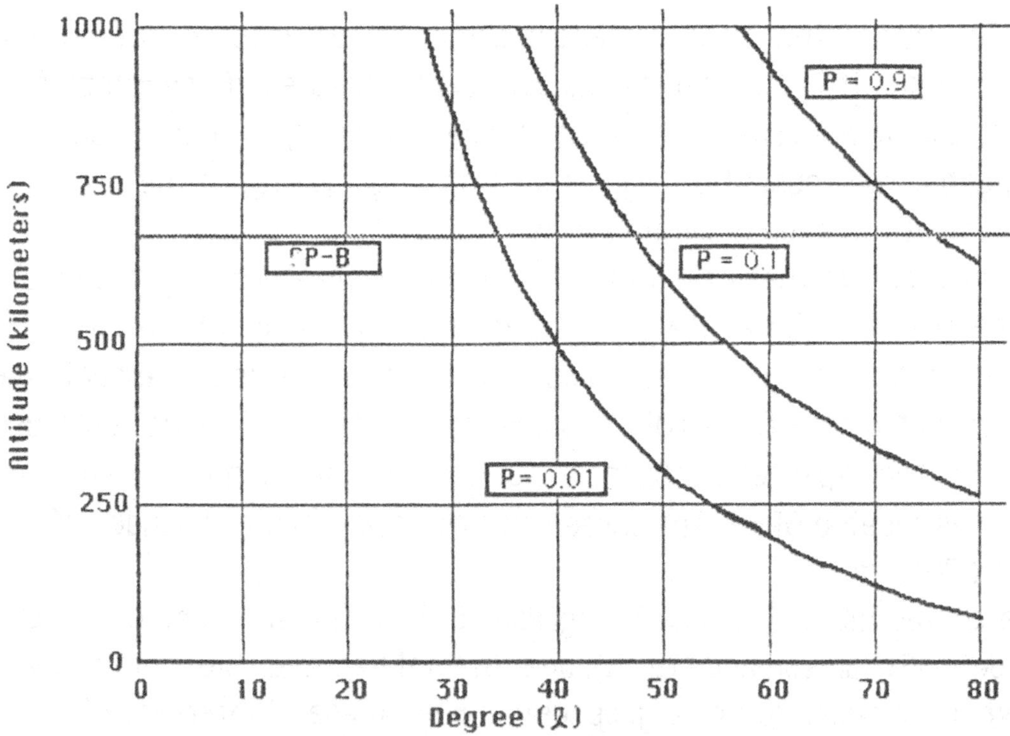

Fig. 3. Contours of Constant Information. $P = {}^{\sigma}J_{lm} / E(J_{lm})$ is the uncertainty expressed as a fraction of the average coefficient magnitude.

curves, the uncertainty is a constant fraction of the expected magnitude of the geopotential coefficients as given in Kaula's rule. Thus, at an altitude of 750km, we expect to produce an uncertainty 90% of the size of the order 70 coefficients. Order 70 is therefore the effective limit of the size of the gravity field GP-B could resolve at that altitude.

The upcoming ESA ARISTOTELES gradiometry mission will permit a solution for the gravity field complementing the GP-B solution very well. We think that the Aristoteles solution would have higher uncertainty than the GP-B solution at orders less than 25, and lower uncertainty at orders above 25. Thus, the nearly simultaneous debut of these two missions would represent a huge leap in the knowledge of the Earth's gravity field.

Figure 4 compares our results to those of Oscar Colombo at the Goddard Space Flight Center and of W. G. Melbourne at the Jet Propulsion Laboratories. The GSFC best-case line is based on the assumption that, as we assumed, the 10-cm phase ambiguity can be resolved; the worst-case line, that it cannot. These curves were produced using other assumptions similar to ours. Colombo performed a covariance analysis to determine the uncertainty in coefficients. GSFC has since sponsored a full-scale simulation of the GP-B mission and data reduction. The results agree very closely with Colombo's covariance analysis. Melbourne used a cylindrical Earth approximation, took the Fourier transform of the potential, and solved the least-squares problem for the reduction in uncertainty in a manner very similar to ours, but in one horizontal dimension instead of two. His results are therefore fairly similar.

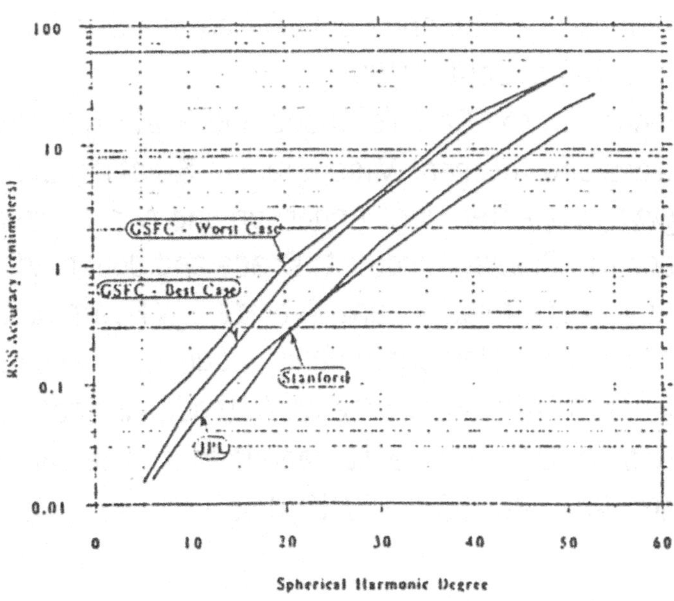

Fig.4. Compared Estimated of Cumulative Geoid Undulation Error

While the flat Earth approximation will not be valid at extremely low orders, the excellent agreement between the three analyses shown in Figure 4 is encouraging. The GSFC uncertainties are higher by a factor of about three in part because the measurement noise was increased by a factor of three to account for the effects of geometric dilution of precision (Colombo)(Navigation) and in part because they assumed only 7 GPS satellites would be visible at any one time instead of 12.

In conclusion, we have shown that the GP-B mission has the capability to drastically reduce uncertainty in the geopotential field of the Earth, especially at low frequencies. The powerful combination of this technique with gradiometry measurements will lead to a much better understanding of the Earth's internal structure and dynamics, as well as many surface phenomena.

REFERENCES

Breakwell, John V., "Satellite Determination of Short Wavelength Gravity Variations", J. of the Astronautical Sciences, Vol. 27 No. 4, Oct.-Dec 1979.

Colombo, O., private communication, 1988.

Eschenbach, Ralph. (Trimble Navigation); private conversation, Sept.1989.

Geometric Dilution of Precision, Navigation, Vol. 25 No.4.

Nerem,R.S., Tapley,B.D., and Shum,C.K.; "A General Ocean Circulation Model Determined in a Simultaneous Solution with the Earth's Gravity Field"; to be published in Proceedings, IAG Symposium 104 on Sea Surface Topography, the Geoid, and Vertical Datums, Sept. 1989.

Peltier,W.R.,"Slow Changes in the Earth's Shape and Gravity Field: Constraints on the Glaciation History and Internal Viscoelastic Stratification"; in *Space Geodesy and Geodynamics*; Allen Joel Anderson and Anny Cazenave, ed.. London, Austin, Tokyo: Academic Press; Harcourt, Brace and Jovanovich, 1986.

Tapley,M.; Breakwell,J.V.; Parkinson,B.; and Everitt,C.W.F.; Impact of the Gravity Probe B Mission on Satellite Navigation and Geodesy, to appear in the Proceedings of the 1989 IAF conference on Astrodynamics.

Williamson, R.G., and Marsh, J.G., Starlette geodyamics: The Earth's Tidal Response, J. Geophys. Res., 90, 9346-9352, 1985.

AN IMPROVED MODEL FOR THE EARTH'S GRAVITY FIELD

C. K. Shum, B. D. Tapley, D. N. Yuan, J. C. Ries and B. E. Schutz
Center for Space Research, The University of Texas at Austin
Austin, Texas 78712 USA

ABSTRACT

Precision orbit determination methods, along with a new technique to compute relative data weights, have been applied to the determination of the Earth's gravity field and other geophysical parameters from the combination of satellite ground based tracking data, satellite altimetry data, and the surface gravimetry data. The University of Texas Earth's gravity field models, PTGF-4 and PTGF-4A, were determined from data sets collected for fifteen satellites, spanning the inclination ranges from 15° to 115°, and surface gravity data. The satellite measurements include laser ranging data, doppler range-rate data, and satellite-to-ocean radar altimeter data, which include the direct height measurement and the differenced measurements at ground track crossings (crossover measurements). The surface gravity data were used in terms of geopotential normal equations (complete to degree and order 50) derived from the Ohio State University $1° \times 1°$ gravity anomaly data. PTGF-4 was computed using satellite tracking data and altimeter crossover data, whereas PTGF-4A was determined using these data sets as well as direct altimeter data and surface gravity data. The estimated parameters for PTGF-4A included geopotential coefficients for a model complete to degree and order 50, tidal coefficients, tracking station coordinates and models for the quasi-stationary sea surface topography. Error analysis and calibration of the formal covariance indicate that GEOSAT orbits can be computed radially at the 15–21 cm level.

INTRODUCTION

At the present time, progress in a number of Earth science research and application areas is limited by the lack of knowledge of the Earth's gravity field. As an example, recent advances in measurement, modeling and conceptual definition in the Earth science community has reached the point where understanding the full three-dimensional structure of the heat and energy transport within the Earth appears possible. The importance of knowledge of the Earth's gravity field in understanding a broad spectrum of phenomena is outlined in a number of national studies, including *A Strategy for Earth Science from Space in the 1980's*, National Research Council Report (1982); *A Comprehensive Program for Solid Earth Science Research*, Earth System Science Committee Report (1986); and *Geophysical and Geodetic Requirements for Global Gravity Field Measurements: 1987–2000*, Report of Gravity Workshop (1987). Planned missions to map the global gravity field include NASA's Gravity Probe B (GP-B) (Everitt et al., 1989), NASA's Geopotential Mission (GRM) (Keating, 1984), the Aristoteles Mission (Bernard and Touboul, 1987), a possible joint ESA-NASA project, and NASA's Superconducting Gravity Gradiometer Mission (SGGM) (Paik et al., 1988). The SGGM aims at recovering the global gravity field to a precision of 2–3 mgal (10–15 cm) with a resolution of 50 km. However, it is not expected that all of these missions will be launched before the year 2000. Consequently, improvement in the near-term will rely on the combination of surface gravity and satellite tracking and altimeter measurements as outlined in the subsequent paragraphs.

Significant progress has been achieved during the last decade in the determination of the spherical harmonic coefficients for the long wavelength components of the Earth's external gravitational potential. A substantial portion of this progress can be directly attributed to the advent of Earth-orbiting artificial satellites and to the ability to observe their motion precisely from either ground-based or satellite-originated tracking data. While the satellite data primarily resolve the long and intermediate wavelengths (≥ 1500 km), surface gravity measurements and the altimeter data are capable of recovering the shorter wavelength components of the Earth's gravity field. Recent trends in gravity model improvement have been driven, in part, by requirements for more accurate satellite orbits to achieve the objectives of the NASA Crustal Dynamics Project (Frey and Bosworth, 1988) and the approved NASA/CNES TOPEX/POSEIDON mission (Stewart et al., 1986). A joint NASA Goddard Space Flight Center (GSFC) and University of Texas Center for Space Research effort to develop an improved model for the Earth's gravity field has been undertaken to meet the orbit accuracy requirement of the TOPEX/POSEIDON mission. The gravity field solution will represent the first complete reiteration of the historical tracking data used to define the NASA GSFC Earth Model series. The GSFC GEM-T1 and GEM-T2 (Marsh et al., 1988; Marsh, 1989), the University of Texas TEG-1 (Tapley et al., 1988), and the PTGF-4 and PTGF-4A gravity fields described in this paper are all preliminary versions of the TOPEX gravity field model.

THEORY AND METHOD

The external gravitational potential, U, due to the Earth's nonspherical mass distribution can be expressed as follows

$$U = \frac{GM}{r} \sum_{l=0}^{\infty} \sum_{m=0}^{l} \left[\frac{R_e}{r} \right]^l \bar{P}_{lm}(\sin\phi) \left[(\bar{C}_{lm} + \Delta\bar{C}_{lm}) \cos m\lambda + (\bar{S}_{lm} + \Delta\bar{S}_{lm}) \sin m\lambda \right] \quad (1)$$

where GM is the product of the gravitational constant and the total mass of the Earth and the atmosphere; R_e is the mean equatorial radius of the Earth; $\bar{P}_{lm}$ are the normalized Legendre associated function of degree l and order m; $\bar{C}_{lm}, \bar{S}_{lm}$ are the the normalized spherical harmonic coefficients whose values are functions of the mass distribution within the Earth and the atmosphere; $\Delta\bar{C}_{lm}, \Delta\bar{S}_{lm}$ are the time-varying components of $\bar{C}_{lm}$ and $\bar{S}_{lm}$ caused by tides; also functions of the tidal coefficients, $C_{lm}^{\pm}$ and $S_{lm}^{\pm}$; and r, ϕ, λ are the Earth-fixed spherical coordinates; r is the radial distance, ϕ is the geocentric latitude and λ is the longitude measured from the Greenwich meridian.

The estimation of $\overline{C}_{lm}, \overline{S}_{lm}, C_{lm}^{\pm}, S_{lm}^{\pm}$ and other orbit and geophysical parameters can be accomplished using a modified least-squares estimation procedure. This estimation procedure, which provides adjustments to satellite orbit-dependent parameters and other geophysical parameters, is given by Tapley (1973) and modified to include the simultaneous estimation of the relative weights for the individual information arrays (Yuan et al., 1988):

$$\hat{x} = (H^T \hat{R}^{-1} H)^{-1} H^T \hat{R}^{-1} y \; ; \; \hat{R}_i = 1/k_i (y_i - H_i \hat{x})^T (y_i - H_i \hat{x}) \cdot I \tag{2}$$

where $\hat{x}$ is the state parameter; $\hat{R}$ is the weighting matrix; I is the identity matrix; H_i is the partial derivative with respect to x for the ith data set; and k_i is the number of observations for the ith dataset.

The system of equations given above can be solved iteratively using orthogonal transformation techniques (Gentleman, 1973). The estimation process has been implemented in the University of Texas Orbit Processor (UTOPIA) software system. The optimal weighting algorithm to combine satellite and nonsatellite information equations was installed in the Large Linear System Solver (LLISS) software system. Vectorized versions of UTOPIA and LLISS are operational on the University of Texas System Center for High Performance Computing Cray X-MP/24 supercomputer. Reference orbits for each of these satellites and information equations for each data set were computed using UTOPIA with *a priori* force and measurement models. The combination solution was performed using LLISS.

DATA AND MODELS

Fifteen satellites were selected for the current gravity model solution. Their orbital characteristics and data types are summarized in Table 1. The inclinations of these satellite orbits vary from 15° for Peole to 115° for GEOS-3. The solution includes data at 90° for OSCAR-14 and NOVA-1. Data types include laser range, one-way range-rate, two-way range-rate, altimeter, altimeter crossover and surface gravity data. The gravity field information equation (complete to degree and order 50) for surface gravity data was computed by Pavlis (1988) using $1° \times 1°$ terrestrial mean gravity anomaly data from Ohio State University and removing signals about degree and order 50 of the geopotential. SEASAT and GEOSAT direct altimeter data were processed into normal points by removing geopotential signals about degree and order 50. The nominal gravitational and nongravitational force models, the Earth orientation and time model, laser, doppler, and direct altimeter measurement models used to compute the reference orbits primarily according to the MERIT Standards (Melbourne et al., 1983) and are summarized by Tapley et al. (1987) and Tapley et al. (1988).

Table 1. Satellite data for the University of Texas gravity models, PTGF-4 and PTGF-4A.

Satellite	Data type/span	Inclination	Eccentricity	Altitude (km)
GEOS-1	Laser (1977, 80-81)	59°	0.072	1600
BE-C	Laser (1977-80, 85-86)	41°	0.026	1130
DI-C	Laser (1971)	40°	0.053	1000
DI-D	Laser (1971)	39°	0.085	1200
OSCAR-14	Doppler (1980)	89°	0.005	1100
GEOS-2	Laser (1977)	106°	0.033	1400
Peole	Laser (1971)	15°	0.015	650
LANDSAT-1	USB doppler (1974)	98°5	0.001	900
GEOS-3	Laser (1976, 80-81)	115°	0.002	830
Starlette	Laser (1983-87)	50°	0.020	900
LAGEOS	Laser (1979-85)	110°	0.004	5900
SEASAT	Laser, Doppler, USB Doppler, Altimeter and Crossover (1978)	108°	0.002	800
NOVA-1	Doppler (1982)	90°	0.002	1200
GEOSAT	Doppler, Altimeter and Crossover (1987)	108°	0.000	800
Ajisai	Laser (1986-87)	50°	0.001	1500

Surface Gravity Data

(50×50) normal equation computed using $1° \times 1°$ terrestrial mean gravity anomaly data from Ohio State University (Pavlis, 1988)

SOLUTION

The list of parameters which are simultaneously estimated with a relative weighting factor for each data set include: (1) geopotential complete to degree and order 50, plus selected coefficients; (2) GM, (3) ocean tides which include long period tides ($m = 0$, $l = 2,3$): Ssa, Sa, Mm, and Mf; diurnal tides ($m = 1$, $l = 2,3,4,5$): Q1, O1, P1, and K1; semi-diurnal tides ($m = 2$, $l = 2,3,4,5$): N2, M2, S2, K2 and T2 ($l = 2$); (4) models of the quasi-stationary sea surface topography for SEASAT and GEOSAT, complete to degree and order 10; (5) altimeter biases for SEASAT and GEOSAT; (6) correction to the bias due to the significant wave height, $H_{1/3}$; (7) doppler and low inclination satellite laser station coordinates; (8) arc parameters for satellite orbits, which include position and velocity vectors, drag and solar radiation pressure coefficients, density correction parameters for selected satellites, and pass-dependent frequency and troposphere biases for doppler satellites. A modified form of Kaula's constraint equation (Kaula, 1966), which was inferred from surface gravity anomaly data, was used as an *a priori* constraint. PTGF-4 is complete to degree and order 50 plus additional resonant geopotential coefficients to degree 54 and order 43, and contains satellite tracking data and altimeter crossover data from SEASAT and GEOSAT. PTGF-4A is complete to degree and order 50 and contains all the data which was included in PTGF-4 as well as direct altimeter data from SEASAT and GEOSAT and surface gravity normal equations supplied by the Ohio State University.

ACCURACY EVALUATION

Efforts to evaluate and calibrate the accuracy of the University Texas gravity models were performed. Tables 2a–2d show comparisons of orbit fits using different gravity models for Starlette, Ajisai, SEASAT and GEOSAT, respectively. The PTGF-4 and PTGF-4A models and the Goddard Earth Models, GEM-T2 and PGS3250 were used for comparison. PGS3520 model was determined using the GEM-T2 data sets as well as the direct altimeter data from SEASAT and the surface gravity anomaly data. Tables 2a and 2b shows that the Starlette and Ajisai five-day orbit fits are at the 11–12 cm level in terms of range residual root-mean-squared (rms) using PTGF-4 and PTGF-4A gravity fields. Table 2c shows that the altimeter crossover and the direct altimeter residual rms for a SEASAT six-day orbit using PTGF-4A is 26 cm and 21 cm respectively. Table 2d indicates that the altimeter crossover and the direct altimeter residual rms for a continuous 17-day GEOSAT orbit using PTGF-4A is 19 cm and 21 cm respectively; indicating that the radial orbit error for GEOSAT is at the 15–20 cm level. Covariance matrices for PTGF-4 and PTGF-4A were calibrated to obtain estimates of errors associated with the gravity field using the consider covariance calibration technique (Yuan et al., 1988). Table 3 shows predicted radial orbit errors using different gravity covariance matrices for TOPEX and GEOSAT. The predicted orbit error for GEOSAT is consistent with the actual orbit fits as shown in Tables 2c and 2d.

Figures 1a and 1b show predicted radial orbit error using the GEM-T2, PTGF-4 and PTGF-4A gravity field versus inclination covariance matrices at the GEOSAT/SEASAT altitude and at the TOPEX altitude, respectively. Figure 1a predicts that the $98°5$ inclination ERS-1 orbit will be accurate to about 70 cm radially for the PTGF-4A model and about 100 cm radially for the GEM-T2 model.

Figures 2a and 2b show the geographically correlated radial orbit error predicted by PTGF-4A covariance matrix for GEOSAT and TOPEX, respectively. Although the rms of the mean gravity error is at the 9 cm level for TOPEX and the 12 cm level for GEOSAT, large errors of the order of 14 cm are still present, for example, in the Pacific basin.

Table 2a. Gravity field accuracy evaluation using Starlette orbit fits.

Five-day orbits $-\bar{r}_o, \bar{v}_o, C_R$, daily C_D, adjusted

Epoch	PTGF-3A (rms)	GEM-T2† (rms)	PGS-3250† (rms)	PTGF-4 (rms)	PTGF-4A (rms)
88/7/24	20	19	16	10	10
88/8/13	17	13	13	10	9
88/8/08	14	15	14	13	13
88/8/13	13	14	14	12	12

· Units for laser are cm; nominal ocean tide model (MERIT) and tracking station coordinates used
† Goddard Space Flight Center gravity models.

Table 2b. Gravity field accuracy evaluation using Ajisai orbit fits.

Five-day and 60-day orbits $-\bar{r}_o, \bar{v}_o, C_R$, 2.5-day C_D, adjusted

Epoch	GEM-T2† (rms)	PGS-3520† (rms)	PTGF-4 (rms)	PTGF-4A (rms)
89/5/01	16	15	13	13
80/5/06	11	11	10	11
89/5/11	16	16	14	14
89/5/16	16	15	12	12
89/4/01*	14	14	12	12

· Units for laser are cm; nominal ocean tide model (MERIT) and tracking station coordinates used
† Goddard Space Flight Center gravity models.
* 60-day orbit with 3-day orbit element corrections to remove long-period signal

Table 2c. Gravity Field Accuracy Evaluation Using SEASAT Orbit Fits.

Six-day orbit – $\bar{r}_o$, $\bar{v}_o$, C_R, daily C_D, adjusted

Epoch		GEM-T2† (rms)	PGS-3520† (rms)	PTGF-4 (rms)	PTGF-4A (rms)
78/9/15	Laser	52	52	46	47
	Doppler	0.69	0.69	0.68	0.72
	S-band	0.44	0.49	0.47	0.47
	Crossover	46	28	34	26
	Altimeter	132	–	134	21

· Altimeter data smoothed to represent gravity spectrum to (50×50)
· Units for doppler data are cm/sec, cm for all other data types
· All data types processed for orbit determination
 † Goddard Space Flight Center gravity models.

Table 2d. Gravity field accuracy evaluation using GEOSAT orbit fits.

17-day orbits – $\bar{r}_o$, $\bar{v}_o$, C_R, daily C_D, density correction parameters adjusted

Epoch		GEM-T2† (rms)	PGS-3520† (rms)	PTGF-4 (rms)	PTGF-4A (rms)
87/1/7	Doppler	0.52	0.52	0.51	0.51
	Crossover	35	23	23	19
	Altimeter	130	–	135	21

· Altimeter data smoothed to represent gravity spectrum to (50×50)
· Units for doppler data are cm/sec, cm for all other data types
· All data types processed for orbit determination
 † Goddard Space Flight Center gravity models.

Table 3. Gravity field accuracy evaluation covariance analysis

GEOSAT orbit: 108° inclination, 800 km altitude

Model	Predicted GEOSAT Radial Orbit Error (cm)	Predicted Crossover Error (cm)
GEM-T1†	54	–
GEM-T2†	25	30
PTGF-4	17	21
PTGF-4A	16	19

TOPEX orbit: 65° inclination, 1335 km altitude

Model	Predicted TOPEX radial orbit error (65° inclination) (cm)
GEM-T1†	25
GEM-T2†	10
PTGF-4	11
PTGF-4A	11

† Goddard Space Flight Center gravity models

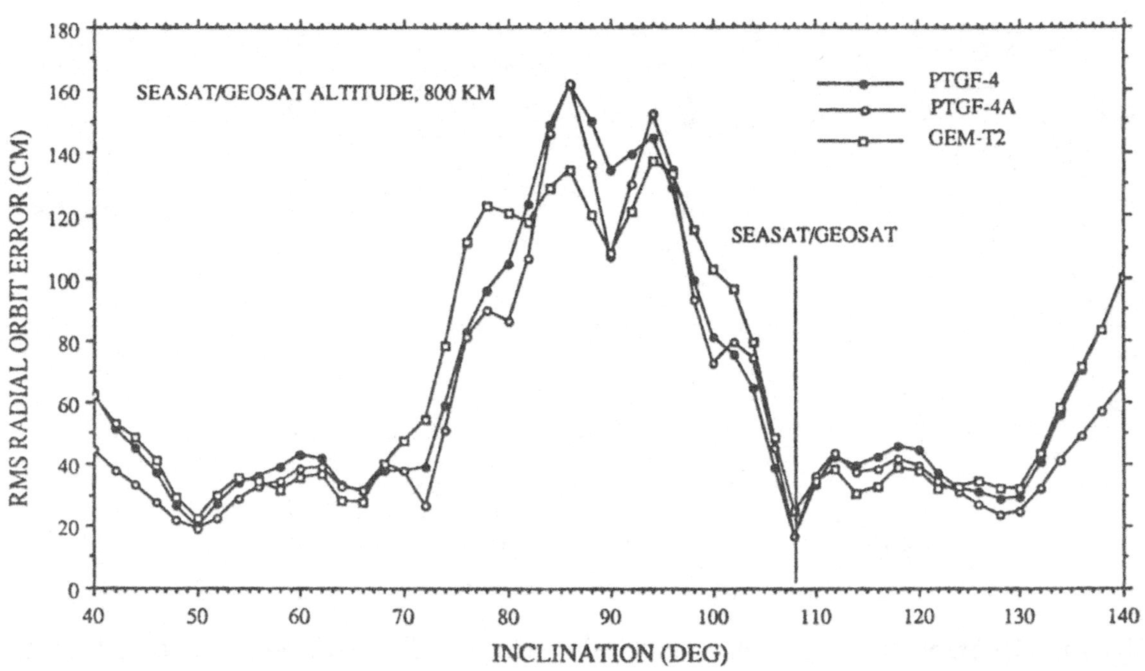

Fig. 1a. Radial orbit error predicted by covariance at SEASAT/GEOSAT altitude.

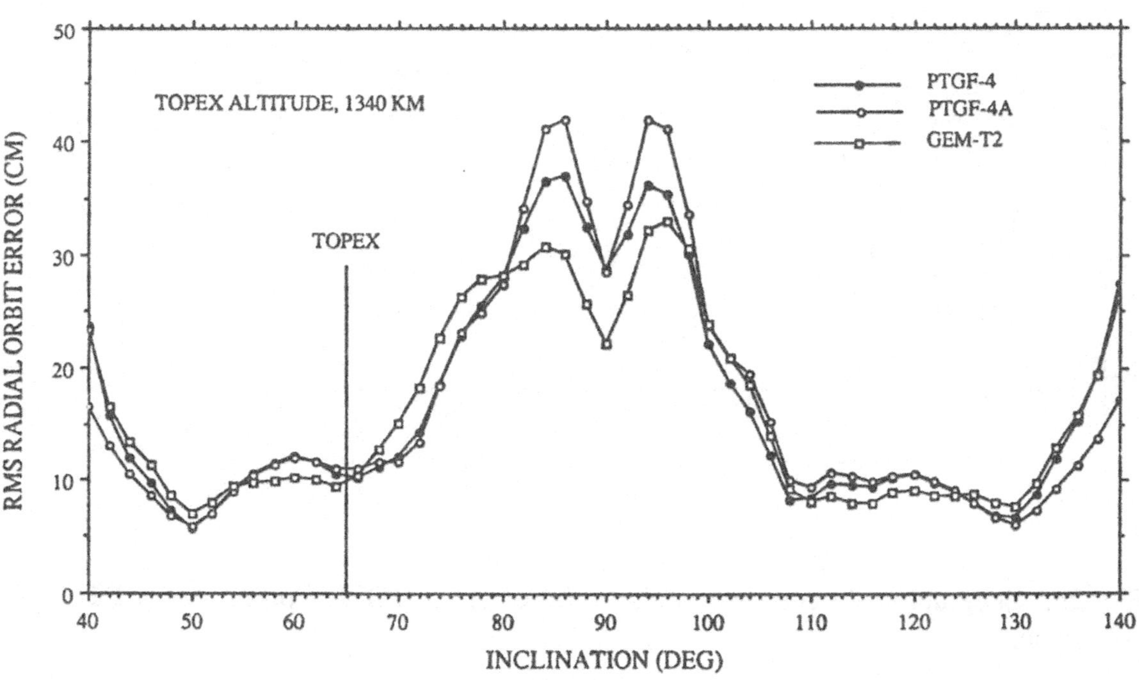

Fig. 1b. Radial orbit error predicted by covariance at TOPEX altitude.

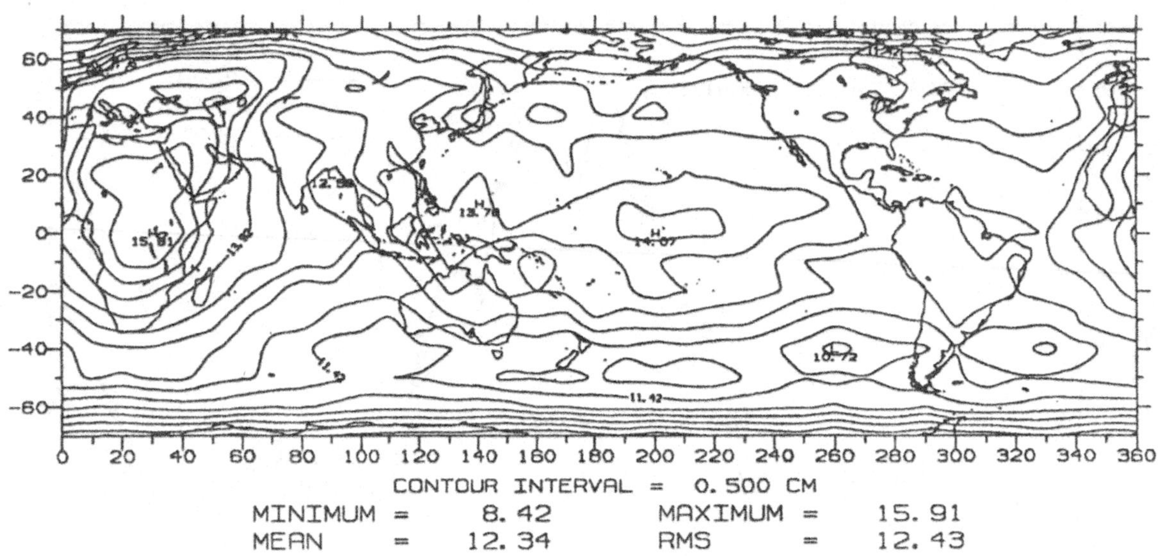

Fig. 2a. Geographically correlated radial orbit error for GEOSAT predicted by PTGF-4A covariance.

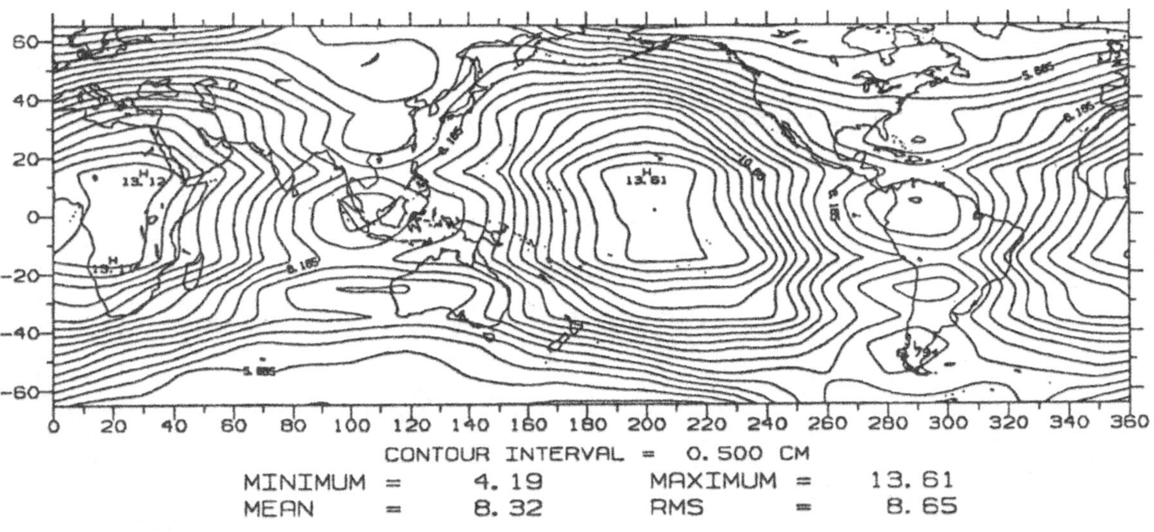

Fig. 2b. Geographically correlated radial orbit error for TOPEX predicted by PTGF-4A covariance.

CONCLUSION

In this investigation, improved models for the Earth's gravity field, PTGF-4 and PTGF-4A were generated. Ground-based tracking data collected for 15 satellites and altimeter crossover data were used to determine the PTGF-4 gravity field model. PTGF-4A contains SEASAT and GEOSAT direct altimeter data and surface gravity data in addition to all the data in PTGF-4. PTGF-4 is complete to degree and order 36 plus selected coefficients. PTGF-4A is complete to degree and order 50. The gravity field models were derived simultaneously with orbit, ocean tides, quasi-stationary sea surface topography, and other geophysical parameters as well as the relative weights for each data set. The fields were evaluated using both data included and data withheld from the solution. Formal covariance matrices were calibrated to reflect realistic error estimates of the gravity models. Evaluations based on orbit fits indicated that GEOSAT orbits can be computed accurately to 15–20 cm radially. Covariance analysis predicts an 11 cm radial accuracy for TOPEX. However, large geographically correlated orbit errors due to the geopotential as large as 14 cm are still present.

Acknowledgments. This research was supported partially by the NASA/Jet Propulsion Laboratory under Contract No. 956689. Additional computing resources were provided by the University of Texas System Center for High Performance Computing.

REFERENCES

Bernard, A. and Touboul, P. (1987). GRADIO: An electrostatic spaceborne gravity gradiometer, *Proceedings of IUGG General Assembly*, Vancouver, BC, Canada.

Everitt, C.W.F., Breakwell, J. V., Tapley, M., DeBra, B. D., Parkinson, B. W., Smith, D. E., Colombo, O., Pavlis, E., Tapley, B. D., Nerem, R. S., Yuan, D. N. and Melbourne, W. G. (1989). Gravity Probe B as a geodesy mission and its implications for TOPEX, *CSTG Bulletin No. 11*.

Frey, H. V. and Bosworth, J. M. (1988). Measuring contemporay crustal motions: NASA's Crustal Dynamics Project, *Earthquakes and Volcanoes* **20**(3).

Gentleman, W. M. (1973). Least square computations by Givens transformation without square roots, *J. Inst. Math. Applic.* **12**.

Kaula, W. M. (1966). Test and combination of satellite determinations of the gravity field with gravimetry, *J. Geophys. Res.* **71**(22), 5304–5314.

Keating, T. (1984). Geopotential Research Mission (GRM), *J. Astron. Sci.,* **32**, 145.

Marsh, J. G. (1989). Earth gravity model computations at the Goddard Space Flight Center, *Eos Trans. AGU* **70**(15), 301.

Marsh, J. G., Lerch, F. J., Putney, B. H., Christodoulidis, D. C., Smith, D. E., Felsentreger, T. L., Sanchez, B. V., Klosko, S. M., Pavlis, E. C., Martin, T. V., Robbins, J. W., Williamson, R. G., Colombo, O. L., Rowlands, D. D., Eddy, W. F., Chandler, N. L., Rachlin, K. E., Patel, G. B., Bhati, S. and Chinn, D. S. (1988). A new gravitational model for the Earth from satellite tracking data: GEM-T1, *J. Geophys. Res.* **93**(B6), 6169–6215.

Melbourne, W., Anderle, R., Feissel, M., King, R., McCarthy, D., Smith, D., Tapley, B. and Vicente, R. (1983). *Project MERIT Standards*, U. S. Naval Observatory,

Circular No. 167, Washington, D. C. 20390.

National Aeronautics and Space Administration, Division of Earth Science and Applications, Geodynamics Branch, (1987). *Geophysical and Geodetic Requirements for Global Gravity Field Measurements, 1987–2000*, Report of a Gravity Workshop, Colorado Springs, Colorado.

National Aeronautics and Space Administration, Office of Space Science and Applications, Earth System Science Committee, Geophysics Panel (1986). *A Comprehensive Program for Solid Earth Science Research, Draft*.

National Research Council, Space Science Board, Committee on Earth Sciences (1982). *A Strategy for Earth Science from Space in the 1980's, Part I: Solid Earth and Oceans*, National Academy Press, Washington, D. C.

Paik, H. J., Leung, J. S., Morgan, S. H. and Parker, J. (1988). Global gravity survey by an orbiting gravity gradiometer, *Eos Trans. AGU* **69**(48).

Pavlis, N., (1988). Modeling and estimation of a low degree geopotential model from terrestrial gravity data, *Report No. 386*, Department of Geodetic Science and Surveying, The Ohio State University, Columbus, Ohio 43210-1247.

Stewart, R. H., Fu, L. L. and Lefebvre, M. (1986). Science opportunities from the TOPEX/POSEIDON Mission, *JPL Publ. 86-18*, NASA Jet Propulsion Laboratory, Pasadena, California.

Tapley, B. D. (1973). Statistical orbit determination theory, *Recent Advances in Dynamical Astronomy*, 396–425, B. D. Tapley and V. Szebehely, Eds., D. Reidel Publ. Co., Holland.

Tapley, B. D., Schutz, B. E., Shum, C. K., Ries, J. C. and Yuan, D. N. (1987). An improved model for the Earth's gravity field, *Proc. IUGG 19th General Assembly*, Vancouver, Canada.

Tapley, B. D., Shum, C. K., Yuan, D. N. and Schutz, B. E. (1988). An improved model for the Earth's gravity field, *Proceedings Chapman Conference on Progress in the Determination of the Earth's Gravity Field*, Ft. Lauderdale, Florida.

Yuan, D. N., Shum, C. K. and Tapley, B. D. (1988). Gravity field determination and error assessment techniques, *Chapman Conference for Progress in the Determination of the Earth's Gravity Field*, Ft. Lauderdale, Florida.

A NEW GEOPOTENTIAL MODEL TAILORED TO GRAVITY DATA IN EUROPE

T. Bašić, H. Denker, P. Knudsen[1], D. Solheim[2], and W. Torge.
Institut für Erdmessung, Universität Hannover (FRG).
[1]) Kort- og Matrikelstyrelsen (Denmark).
[2]) Statens Kartverk (Norway).

ABSTRACT

The European north-south GPS traverse was established to control and improve the European geoid. Comparisons of GPS and levelling data with different geoid computations show a strong slope (1.5 m/700 km) between Denmark and Norway. This slope was suspected to be caused by bad quality of the up to now available $6' \times 10'$ gravity values for Scandinavia. Therefore an updated set of $0.5° \times 0.5°$ mean free-air gravity anomalies was calculated for Scandinavia. The new geopotential model (IFE88E2) has been calculated using this new set of $0.5° \times 0.5°$ mean free-air gravity anomalies merged with other available $0.5° \times 0.5°$ mean values for Europe. The calculation of the tailored model, complete to degree and order 360, is based on the OSU86F coefficient set, that was used as a start model. The RMS value of the differences between the 6715 mean values and the model derived values decreased from 15.8 mgal to 6.0 mgal for OSU86F resp. IFE88E2. The magnitude of residual point free-air anomalies relative to OSU86F and IFE88E2 were evaluated in 1495 points in Scandinavia. The RMS values of the differences were 23.4 mgal and 18.5 mgal respectively. Especially comparisons of the two models with GPS/levelling data in Europe show an improved accuracy of the IFE88E2 model. The RMS value of the differences relative to OSU86F is 0.774 m and decreases to 0.322 m for IFE88E2. Furthermore, the slope between Denmark and Norway has almost disappeared.

INTRODUCTION

High degree spherical harmonic expansions of the Earth's gravitational potential are available for about one decade (eg. Rapp, 1978). These harmonic expansions have proven to be very useful for local gravity field determinations as demonstrated e.g. by Tscherning and Forsberg (1986) or Denker (1988). However, as progress in the collection and compilation of new gravity field data is slow, improved geopotential models are released only at time intervals of approximately five years. In the case that the available global models (e.g. Rapp and Cruz, 1986, Wenzel, 1985) do not approximate the regionally existing gravity field data very well, large data collection areas have to be used in order to get good local gravity field solutions (see e.g. Forsberg and Solheim, 1988). However, this makes computations much more cumbersome and expensive, especially the application of collocation techniques might even be impossible. Therefore, one convenient way is the improvement of an available geopotential model by regional gravity field data, that can then be used for local computations in connection with a small data collection area. Since this strategy has proven to yield very good results for local computations (Denker, 1988, and Bašić, 1989), we present in this paper a new geopotential model (IFE88E2) that has been tailored to an updated set of gravity anomalies for Europe. However, the major initiative to compute a new tailored model came from the comparisons of previous geoid solutions with GPS and levelling results from the European GPS traverse, showing a strong slope in the central part that was suspected to be caused by bad quality of the gravity data in this area (Torge et al., 1989). The new tailored model is compared with results from the European GPS traverse, and several other tests have been performed.

THE METHOD

The standard representation of the Earth's disturbing potential is taken as follows:

$$T(r,\theta,\lambda) = \frac{GM}{r}\sum_{l=2}^{\infty}\left(\frac{a}{r}\right)^{l}\sum_{m=0}^{l}\left(\Delta\overline{C}_{lm}cosm\lambda + \Delta\overline{S}_{lm}sinm\lambda\right)\overline{P}_{lm}(cos\theta) \qquad (1)$$

where

r,θ,λ are the polar coordinates of the point at which T is to be determined,

GM is the geocentric gravitational constant,

a is the semimajor axis of the reference ellipsoid,

$\overline{P}_{lm}$ are the fully normalized associated Legendre functions, and

$\Delta\overline{C}_{lm},\Delta\overline{S}_{lm}$ are the fully normalized spherical harmonic coefficients of the disturbing potential.

In spherical approximation the expansion for gravity anomalies becomes

$$\Delta g(r,\theta,\lambda) = \frac{GM}{r^2}\sum_{l=2}^{\infty}(l-1)\left(\frac{a}{r}\right)^{l}\sum_{m=0}^{l}\left(\Delta\overline{C}_{lm}cosm\lambda + \Delta\overline{S}_{lm}sinm\lambda\right)\overline{P}_{lm}(cos\theta). \qquad (2)$$

However, the use of ellipsoidal approximations is not difficult and is treated in detail by Wenzel (1985) and Pavlis (1988).

As we are primarily interested to find the potential coefficients from gravity anomalies, the orthogonality relationships can be applied on (2) yielding in

$$\begin{Bmatrix}\Delta\overline{C}_{lm}\\ \Delta\overline{S}_{lm}\end{Bmatrix} = \frac{1}{4\pi}\int\int_{\sigma}\frac{r^2}{GM}\left(\frac{r}{a}\right)^{l}\frac{1}{l-1}\Delta g(r,\theta,\lambda)\begin{Bmatrix}cosm\lambda\\ sinm\lambda\end{Bmatrix}\overline{P}_{lm}(cos\theta)d\sigma \qquad (3)$$

where σ is the unit sphere.

The actual evaluation of (3) is carried out using a set of mean gravity anomalies. A mean anomaly can be computed from (2) as follows (Wenzel, 1985):

$$\overline{\Delta g} = \frac{GM}{r^2\Delta\sigma}\sum_{l=2}^{\infty}(l-1)\left(\frac{a}{r}\right)^{l}\sum_{m=0}^{l}\int_{\lambda_W}^{\lambda_E}(\Delta\overline{C}_{lm}cosm\lambda + \Delta\overline{S}_{lm}sinm\lambda)d\lambda\int_{\theta_N}^{\theta_S}\overline{P}_{lm}(cos\theta)sin\theta d\theta \qquad (4)$$

where $\Delta\sigma$ is the surface element on the unit sphere, and $\theta_N,\theta_S,\lambda_W,\lambda_E$ are the boundaries of the integration area.

After describing the underlaying theoretical background we will now discuss in more detail the case, that a global geopotential model is to be improved by regional data. The discussion follows to a great extent the ideas published by Weber and Zomorrodian (1988) and refined by Bašić (1989).

According to Weber and Zomorrodian (1988), the mean gravity anomalies $\overline{\Delta g'}$, computed from the potential coefficients of the start model $\Delta C'_{lm},\Delta S'_{lm}$ using (4), are subtracted from the terrestrial mean gravity anomalies $\overline{\Delta g}$ yielding differences:

$$\delta\overline{\Delta g'} = \overline{\Delta g} - \overline{\Delta g'} \qquad (5)$$

These residual anomalies can then be expanded in spherical harmonics using the following expression

$$\left\{ \begin{matrix} \delta\Delta\overline{C'}_{lm} \\ \delta\Delta\overline{S'}_{lm} \end{matrix} \right\} = \frac{1}{4\pi} \sum_{i=1}^{n} \frac{r_i^2}{GM} \left(\frac{r_i}{a}\right)^l \frac{1}{l-1} \frac{1}{\beta_l} \delta\overline{\Delta g_i'} \int\int_{\Delta\sigma_i} \left\{ \begin{matrix} cosm\lambda \\ sinm\lambda \end{matrix} \right\} \overline{P}_{lm}(cos\theta)d\sigma \qquad (6)$$

where n is the number of differences occurring between model and terrestrial anomalies, $\Delta\sigma$ is the area of integration, and β_l is the Pellinen damping function. The β_l function can be viewed as a de-smoothing operator that tries to take into account that frequencies are damped out in taking the average to obtain the mean anomaly. However, it should be noted here, that in our case the β_l damping function is only applied for the residual anomalies $\delta\overline{\Delta g'}$ and the respective potential difference coefficients. The β_l function can be computed using the following expression

$$\beta_l = \frac{1}{1 - cos\psi_0} \frac{1}{\sqrt{2l+1}} (P_{l-1}(cos\psi_0) - P_{l+1}(cos\psi_0)) \qquad (7)$$

where ψ_0 is the radius of a spherical cap with the same size as the area of integration $\Delta\sigma$. A recurrence procedure for the computation of the β_l can be found in Sjoeberg (1980).

The numerical evaluation of (6) does not cause any problems since efficient algorithms are at hand for the integration of $cosm\lambda$ and $sinm\lambda$, and recurrence relations are available for the integrated Legendre functions (see e.g. Paul, 1978, and Wenzel, 1985). Moreover, for the evaluation of formula (6) it is also possible to apply the very efficient procedure developped by Colombo (1981), that takes advantage of FFT techniques. In this case, the n given $\delta\Delta g_i'$ values have to be complemented with zero values to form a complete set on a sphere (see also Pavlis, 1988). This is also done implicitly in equation (6), and therefore both procedures will yield identical results.

Finally, the coefficients of the modified potential model are obtained by adding the potential difference coefficients obtained from (6) to the start coefficients:

$$\left\{ \begin{matrix} \Delta\overline{C''}_{lm} \\ \Delta\overline{S''}_{lm} \end{matrix} \right\} = \left\{ \begin{matrix} \Delta\overline{C'}_{lm} \\ \Delta\overline{S'}_{lm} \end{matrix} \right\} + \left\{ \begin{matrix} \delta\Delta\overline{C'}_{lm} \\ \delta\Delta\overline{S'}_{lm} \end{matrix} \right\} . \qquad (8)$$

Mean anomalies from this improved set of potential coefficients can be obtained in analogy to (4):

$$\overline{\Delta g''} = \frac{GM}{r^2\Delta\sigma} \sum_{l=2}^{l_{max}} (l-1)\left(\frac{a}{r}\right)^l \sum_{m=0}^{l} \int_{\lambda_W}^{\lambda_E} (\Delta\overline{C''}_{lm} cosm\lambda + \Delta\overline{S''}_{lm} sinm\lambda)d\lambda \int_{\theta_N}^{\theta_S} \overline{P}_{lm}(cos\theta)sin\theta d\theta$$

$$(9)$$

where l_{max} is the maximum degree of the harmonic expansion. Then, the differences $\delta\overline{\Delta g''}$ will give the residual misfit to the new model where

$$\delta\overline{\Delta g''} = \overline{\Delta g} - \overline{\Delta g''} \qquad (10)$$

is defined in analogy to (5).

However, due to the limited degree of the expansion (9) and due to the approximation through (7), the differences $\delta\overline{\Delta g''}$ will not vanish completely. Formula (6) to (10) can be used iteratively until no further significant decrease of $\overline{\delta \Delta g''}$ exists (Weber and Zomorrodian, 1988).

THE DATA

A new set of $0.5° \times 0.5°$ mean gravity anomalies was constructed for Scandinavia using the gravity data base established at the Danish Geodetic Institute (now: Kort- og Matrikelstyrelsen, i.e. National Survey and Cadastre). For the construction of the new mean anomaly set for Scandinavia, the area $54° \leq \phi \leq 81.5°$, $3° \leq \lambda \leq 36°$ (including the Barents Sea Svalbard) was divided into small grid cells of $2.5' \times 5'$. For each of these cells one and only one gravity anomaly observation was selected. The estimated errors were used as the main criteria for the selection of the observations and the distances to the center of the cells were used as a second criteria.

The contribution from GPM2 (Wenzel, 1985) was then subtracted from the selected observations, and anomalies were estimated in a $5' \times 10'$ grid by collocation. Finally, the contribution from GPM2 was restored, while new $0.5° \times 0.5°$ mean free-air anomalies were obtained by taking simple averages. This procedure provided a mean anomaly for each $0.5° \times 0.5°$ block covering the above defined area. For the final product mean anomalies were deleted where no observation was located inside the corresponding $0.5° \times 0.5°$ block. This editing resulted in a set of 2726 mean free-air anomalies for Scandinavia.

The second source of gravity data, that was used in this study, is the $6' \times 10'$ mean gravity data base established at Institut für Erdmessung (IFE). The compilation of this data set, covering large parts of Europe, is described in detail by Weber (1984). From this data set, $0.5° \times 0.5°$ mean anomalies were computed by taking simple averages, if at least ten $6' \times 10'$ values were available inside a $0.5° \times 0.5°$ block. Using this procedure, 5569 $0.5° \times 0.5°$ mean anomalies were computed from the IFE source. These mean anomalies were subsequently replaced or completed by the Scandinavian mean anomalies where these values were available. This merging, using 5569 values derived from the IFE source and the 2726 Scandinavian values, has provided a set of 6715 $0.5° \times 0.5°$ mean gravity anomalies for Europe that were used for the computation of the tailored model IFE88E2.

The Scandinavian $0.5° \times 0.5°$ mean anomalies and the IFE mean values have 1580 common values. The mean value and the RMS value of the differences between both sources is 0.22 mgal and 8.73 mgal respectively. The minimum and maximum differences are -37.41 mgal and 58.90 mgal respectively. A contour plot of the differences is shown in Figure 1. All anomalies are referring to the Geodetic Reference System 1980.

THE MODEL DEVELOPMENT

The merged $0.5° \times 0.5°$ mean anomaly data set, described in the previous section, was used to compute the new tailored model IFE88E2. Since the goal was to develop a tailored model complete to degree and order 360, the obvious choice for the start model is a field which is also complete to degree and order 360. For this reason the geopotential model OSU86F (Rapp and Cruz, 1986) was choosen as a start model.

The differences $\overline{\delta \Delta g'}$ between corresponding $0.5° \times 0.5°$ mean anomalies from the merged terrestrial data and the start model OSU86F were calculated using equation (5) and then expanded into a series of spherical harmonics up to degree 360 according to equation (6). The low degree coefficients of OSU86F were not modified due to the limited data collection area and due to the fact that these coefficients are well determined from the analysis of satellite orbit perturbations. Extensive numerical investigations performed by Bašić (1989) have shown that it is advantageous to taper off the coefficient changes for the low degrees in equation (6) to reduce leakage effects. This was achieved by multiplying the potential difference coefficients, obtained from (6), by a weight function p_l depending on the harmonic degree l. Here, a weight function being zero up to degree 7, increasing linearly from zero to 1.0 between degree 7 and 20, and being 1.0 above degree 20, was used (see Figure 2). This procedure is very similar to techniques used by Torge et al. (1982) in connection with the least squares spectral combination method. A consequence, however, of using such a procedure is a reduced rate of convergence of the whole process and the necessity to carry out the computations iteratively (Bašić, 1989). In total, 5 iterations were performed for the computation of the tailored model IFE88E2.

Results from the comparison of the $0.5° \times 0.5°$ mean gravity anomalies with OSU86F as well as IFE88E2 are given in Table 1. As can be seen from Table 1, the RMS difference between the terrestrial mean gravity anomalies and the start model OSU86F amounts to 15.8 mgal and reduces by almost a factor 3 to 6.0 mgal for the tailored model IFE88E2. Thus, the tailored model IFE88E2 fits the gravity anomalies in Europe significantly better than OSU86F. Independent comparisons presented in the next section show that also the overall behaviour of the tailored model is better than that of OSU86F.

Table 1. Comparison of Terrestrial $0.5° \times 0.5°$ Mean Gravity Anomalies with OSU86F and IFE88E2.

	OSU86F	IFE88E2
Number	6715	6715
Bias	0.16 mgal	0.01 mgal
Std.dev.	15.77 mgal	5.98 mgal
RMS	15.77 mgal	5.98 mgal
Min.	-97.44 mgal	-62.76 mgal
Max.	139.78 mgal	43.29 mgal

Anomaly degree variances were computed for both the start model OSU86F and the tailored model IFE88E2 using the formula

$$\sigma_l^2(\Delta g) = \left(\frac{GM(l-1)}{R_E^2} \right)^2 \left(\frac{a}{R_E} \right)^{2l} \sum_{m=0}^{l} \left(\Delta C_{lm}^2 + \Delta S_{lm}^2 \right) \tag{11}$$

where R_E is the mean Earth radius. The resulting degree variances for both models are shown in Figure 3. In additon, the start model OSU86F and the tailored model IFE88E2 were compared by computing degree variances of the coefficient differences using the formula

$$\delta_l^2(\Delta g) = \left(\frac{GM(l-1)}{R_E^2} \right)^2 \left(\frac{a}{R_E} \right)^{2l} \sum_{m=0}^{l} \left\{ \left(\Delta C_{lm}^{OSU86F} - \Delta C_{lm}^{IFE88E2} \right)^2 + \left(\Delta S_{lm}^{OSU86F} - \Delta S_{lm}^{IFE88E2} \right)^2 \right\}.$$

$$\tag{12}$$

The result of this comparison is also shown in Figure 3. The magnitude of the anomaly degree variances associated with the tailored model IFE88E2 has only increased slightly relative to the start model OSU86F. The anomaly variances are 723 and 730 mgal^2 respectively. The anomaly variance of the coefficient differences is only 3 mgal^2. However, these numbers are to be considered globally according to their definition.

EVALUATION OF THE TAILORED MODEL IFE88E2

As a first test, the start model OSU86F and the tailored model IFE88E2 were compared with GPS and levelling results from the European GPS traverse, described in Torge et al. (1989). The locations of the GPS stations are shown in Figure 4. The results of the GPS/levelling comparisons, after subtracting a common bias, are shown in Figure 5. The RMS discrepancy drops from 0.77 m for OSU86F to 0.32 m for IFE88E2, thus proving a significant improvement of IFE88E2. The tailored model shows a very good long wavelength quality and behaves clearly superior as compared with OSU86F and most other geopotential models discussed in Torge et al. (1989). The slope of about 1.5 m over 700 km between Denmark and southern Norway, visible in OSU86F and the other models evaluated in Torge et al. (1989), has disappeared. This is a strong indication that systematic errors in the former Scandinavian mean anomaly set are responsible for the slope visible in all earlier geopotential models. Noticeable is also the improvement in the southern part of the traverse section, although both models are using essentially the same gravity data in this area. However, there are some spikes visible in northern Norway, which are probably due to the rough topography in that area.

The model was also tested against point gravity observations. In Scandinavia a comparison of 1495 observations with OSU86F and IFE88E2 resulted in the mean values and the standard deviations listed in Table 2. A similar comparison was carried out by Tscherning (1989) in a $0.75° \times 1.00°$ area covering the Strait of Gibraltar (see Table 3).

Table 2. Comparison of Point Gravity Observations with OSU86F and IFE88E2 in Scandinavia.

	Free-air Anomalies	Anomalies Relative to OSU86F	Anomalies Relative to IFE88E2
Number	1495	1495	1495
Mean	-0.22 mgal	-1.45 mgal	-1.58 mgal
Std.Dev.	30.26 mgal	23.40 mgal	18.43 mgal

Table 3. Comparison of Point Gravity Observations with OSU86F and IFE88E2 in the Strait of Gibraltar Area.

	Free-air Anomalies	Anomalies Relative to OSU86F	Anomalies Relative to IFE88E2
Number	2170	2170	2170
Mean	-19.9 mgal	-5.0 mgal	-1.8 mgal
Std.dev.	43.7 mgal	29.6 mgal	25.9 mgal

The results listed in Table 2–3 show that the tailored model IFE88E2 fits the gravity observations better than the start model OSU86F. In both cases the standard deviations have decreased by 4–5 mgal, and the offset in the Strait of Gibraltar area has decreased from 5 to 1.8 mgal. Also in this case, the improvement in the Strait of Gibraltar area is noticeable, as probably the same gravity data have been used in the computation of both models.

CONCLUSIONS

A new geopotential model, complete to harmonic degree and order 360, was tailored to gravity data in Europe including an updated set of mean anomalies for Scandinavia. Compared with the start model OSU86F, the tailored model IFE88E2 shows a significant improvement. Especially comparisons of the two models with GPS/levelling data from the European GPS traverse show an improved accuracy of the IFE88E2 model. The RMS value of the differences relative to OSU86F is 0.774 m and decreases to 0.322 m for the tailored model IFE88E2. Thus, the computation of tailored models is an efficient and convenient way to take new gravity field data into account if existing models do not appear to fit these data well.

Acknowledgement. We would like to acknowledge Prof. Dr.-Ing. H.-G. Wenzel and Dr.-Ing. G. Weber for providing the basic software and helpful discussions.

REFERENCES

Bašić, T. (1989). Untersuchungen zur regionalen Geoidbestimmung mit "dm"–Genauigkeit. Wiss. Arb. Fachrichtung Vermessungswesen der Universität Hannover, No. 157, Hannover.

Colombo, O. (1981). Numerical Methods for Harmonic Analysis on the Sphere. Dept. of Geodetic Science and Surveying, Rep. No. 310, The Ohio State University, Columbus/Ohio.

Denker, H. (1988). Hochauflösende regionale Schwerefeldbestimmung mit gravimetrischen und topographischen Daten. Wiss. Arb. Fachrichtung Vermessungswesen der Universität Hannover, No. 156, Hannover.

Forsberg, R., D. Solheim (1988). Performance of FFT Methods in Local Gravity Field Modelling. Presented Chapman Conference on Progress in the Determination of the Earth's Gravity Field, Fort Lauderdale/Florida.

Paul, M.K. (1978). Recurrence Relations for Integrals of Associated Legendre Functions. Bulletin Géodésique 52, 177–190.

Pavlis, N. (1988). Modeling and Estimation of a Low Degree Geopotential Model From Terrestrial Gravity Data. Dept. of Geodetic Science and Surveying, Rep. No. 386, The Ohio State University, Columbus/Ohio.

Rapp, R.H. (1978). A Global $1° \times 1°$ Anomaly Field Combining Satellite, Geos-3 Altimeter, and Terrestrial Data. Dept. of Geodetic Science and Surveying, Rep. No. 278, The Ohio State University, Columbus/Ohio.

Rapp, R.H., J.Y. Cruz (1986). Spherical Harmonic Expansion of the Earth's Gravitational Potential Field to Degree 360 Using $30'$ Mean Anomalies. Dept. of Geodetic Science and Surveying, Rep. No. 376, The Ohio State University, Columbus/Ohio.

Sjoeberg, L. (1980). A Recurrence Relation for the β_l Function. Bulletin Géodésique 54, 67-72.

Torge, W., G. Weber, H.-G. Wenzel (1982). Computation of a High Resolution European Gravimetric Geoid (EGG1). Proceedings of the 2nd Symposium on the Geoid in Europe and the Mediterranean Area, Rome, 437-460, Instituto Militare Geografico Militare Italiano, Florence.

Torge, W., T. Bašić, H. Denker, J. Doliff, H.-G. Wenzel (1989). Long Range Geoid Control through the European GPS Traverse. Deutsche Geodätische Kommission, Reihe B, No. 290, München.

Tscherning, C.C., R. Forsberg (1986). Geoid Prediction in the Nordic Countries From Gravity and Height Data. Boll. Geod. Sci. Aff. 46, 21-43.

Tscherning, C.C. (1989). Achievable Accuracy for Geoid Height Differences Across the Strait of Gibraltar. Presented First International Workshop on Geodesy for the Europe–Africa Fixed Link Feasability Studies in the Strait of Gibraltar, Madrid, March 8–10, 1989.

Weber, G. (1984). Hochauflösende Freiluftanomalien und gravimetrische Lotabweichungen für Europa. Wiss. Arb. Fachrichtung Vermessungswesen der Univeristät Hannover, No. 135, Hannover.

Weber, G., H. Zomorrodian (1988). Regional Geopotential Model Improvement for the Iranian Geoid Determination. Bulletin Géodésique 62, 125-141.

Wenzel, H.-G. (185). Hochauflösende Kugelfunktionsmodelle für das Gravitationspotential der Erde. Wiss. Arb. Fachrichtung Vermessungswesen der Universität Hannover, No. 137, Hannover.

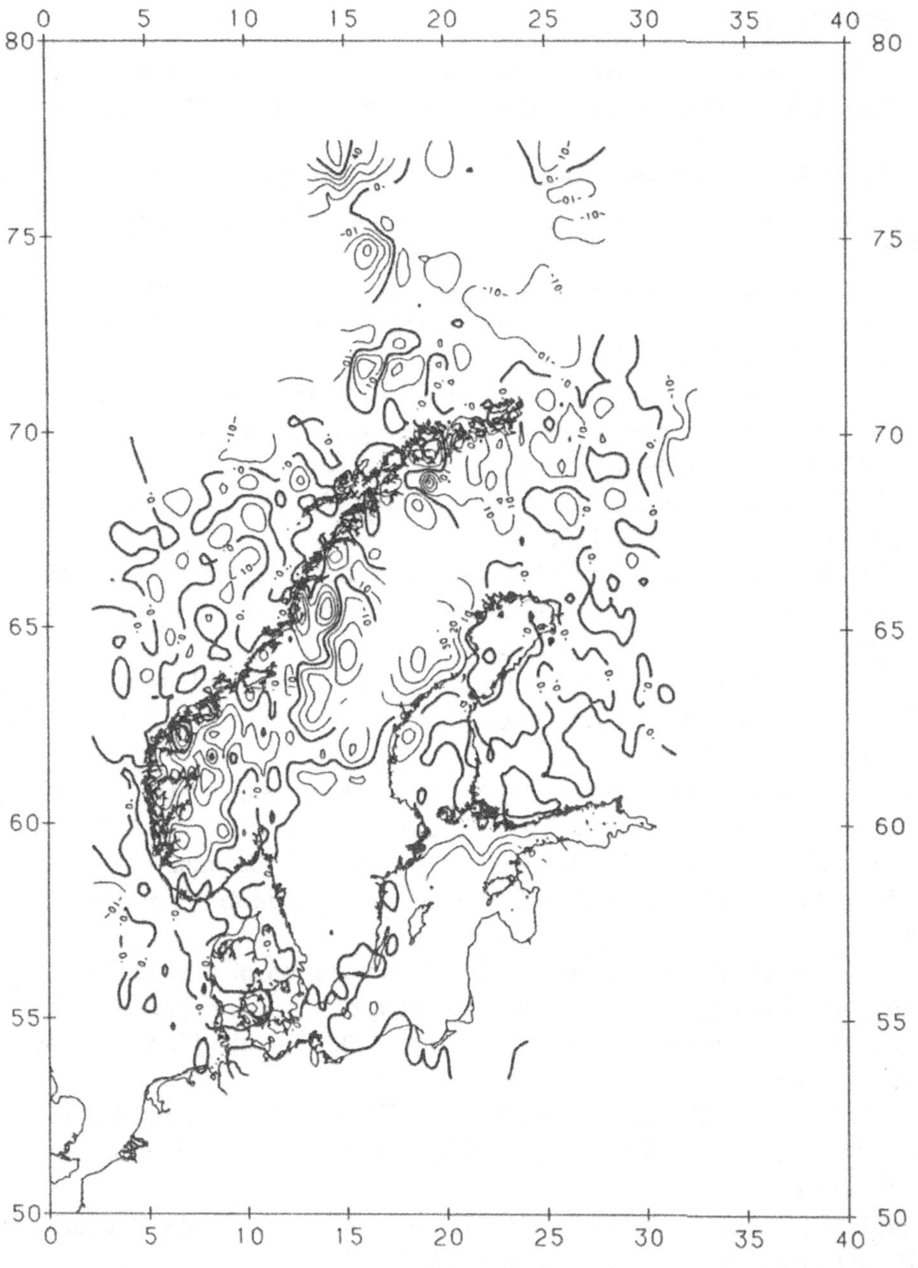

Figure 1. Differences between the Scandinavian Mean Values and the IFE Mean Values in 1580 Common Blocks (C.I. 10 mgal).

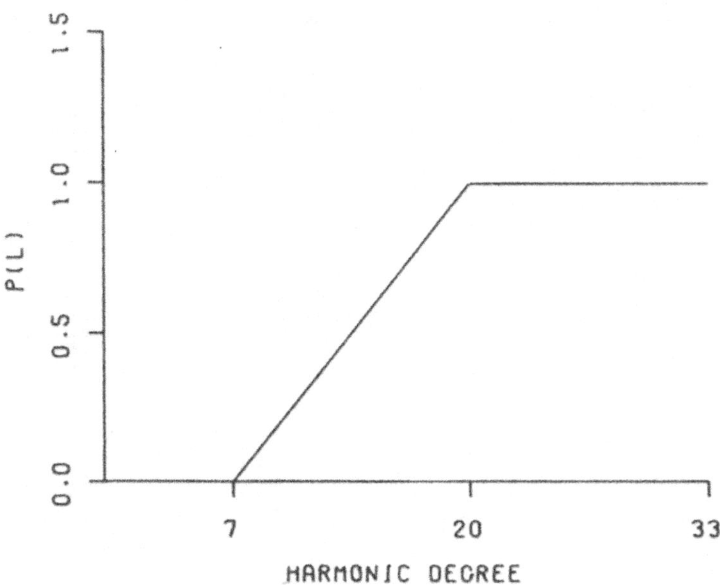

Figure 2. Weight Function p_l

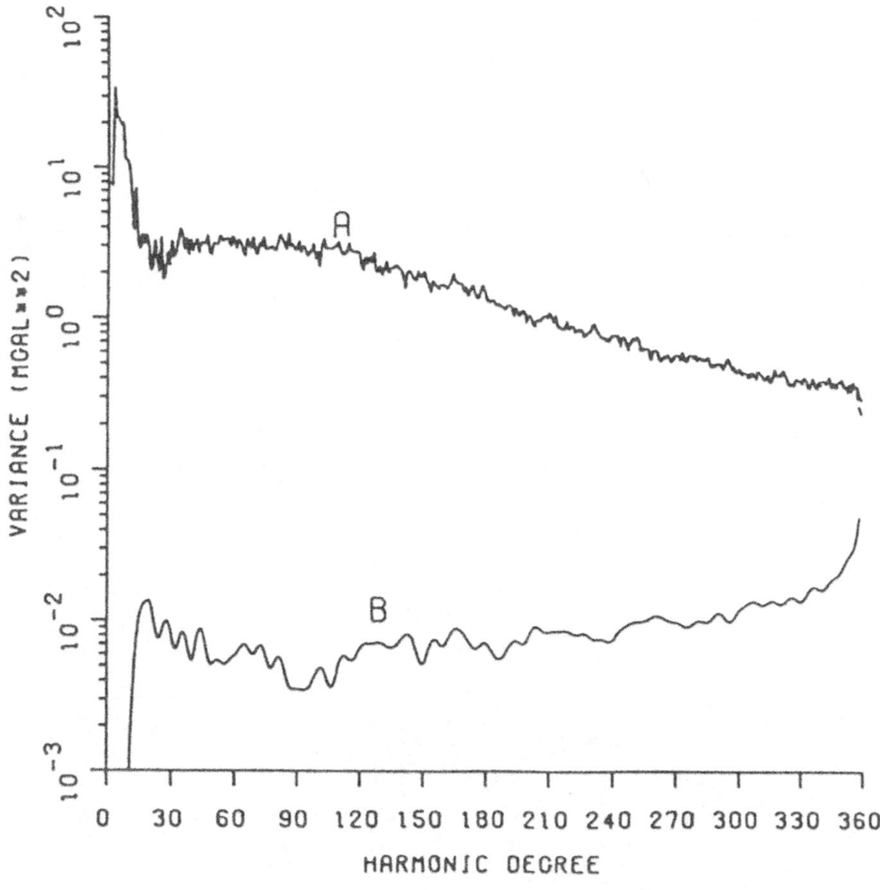

Figure 3. A: Anomaly Degree Variances of OSU86F (Dashed Line) and IFE88E2 (Solid Line). B: Anomaly Degree Variances of Coefficient Differences.

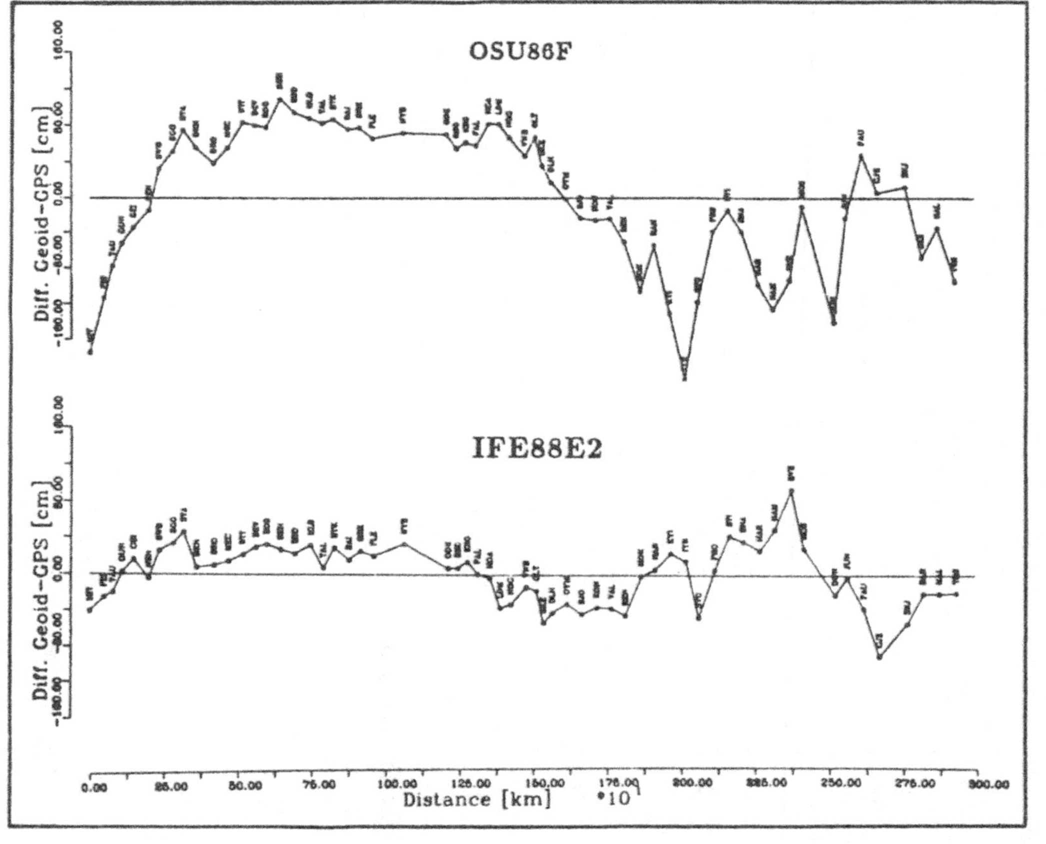

Figure 4. European North-South GPS Traverse.

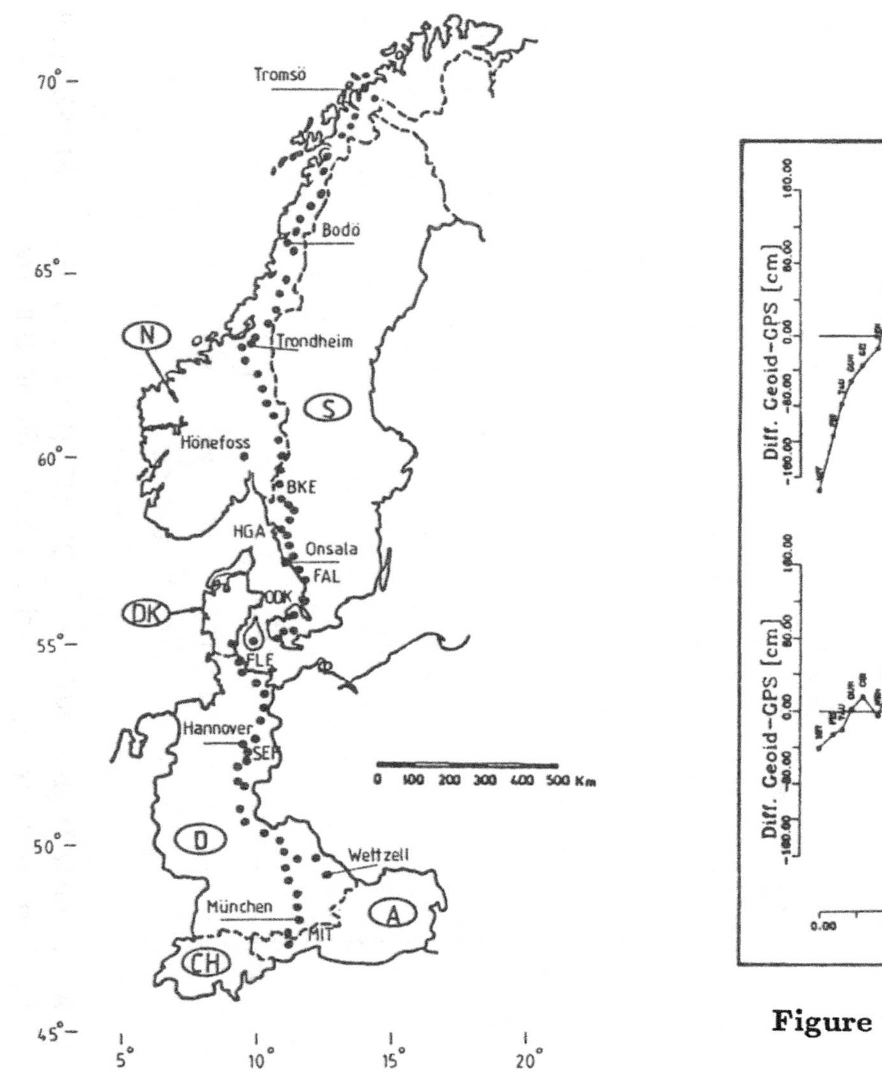

Figure 5. Comparison of GPS/Levelling with OSU86F and IFE88E2.

GRAVITY FIELD OF PHOBOS AND ITS LONG TERM VARIATIONS

M. Burša
Astronomical Institute, Czechoslovak Academy of Sciences,
Budečská 6, 120 23 Praha 2, Czechoslovakia

Z. Martinec, K. Pěč
Charles University
V Holešovičkách 2, 180 00 Praha 8, Czechoslovakia

INTRODUCTION

The only information of the Phobos gravitational field is
the data given by spacecrafts imaging the Phobos surface.
The first detailed model of the Phobos boundary topography
was established by Turner (1978), based on the Mariner 9
television pictures. Provided that Phobos is a homogeneous
body, Sagitov et al. (1981), evaluated the Stokes parameters of the external gravitational field by numerical integration. An independent analytical approach by Martinec et
al. (1989) using the spherical harmonic expansion of the
Phobos topography checked the earlier results and, in fact,
proved Sagitov's (1981) results. Unfortunately, due to many
possible shortcomings of the Turner's globe, results by Sagitov et al. (1981), as well as, by Burša et al. (1988)
cannot be used for the correct representation of the Phobos
external gravitational field.
   The latest global control network of 98 surface features
on Phobos was established in Viking Orbiter images (Duxbury
and Callahan, 1989). Unfortunately, the Mariner 9 data was
not included. Only 98 control points available limited the
maximum degree and order of harmonic expansion of the Phobos topography up to six (Duxbury, 1989).

GRAVITATIONAL FIELD OF PHOBOS

For a homogeneous body, there is a direct analytical relationship between the topography and external gravitational
field harmonic coefficients (Martinec et al., 1989):

$$A_{jm} = \frac{3}{(j+3)(2j+1)} \; \frac{E_{jm}^{(j+3)}}{1 + E_{00}^{(3)}/2\sqrt{\pi}} \; ; \qquad (1)$$

A_{jm} are the Stokes parameters (potential coefficients) of the gravitational field and $E_{jm}^{(j+3)}$ are the power topography coefficients. Nevertheless, this relationship is non-linear if the power topography coefficients $E_{jm}^{(j+3)}$ are expressed in terms of the topography coefficients E_{jm} (Martinec et al., 1989). Only in the linearized case when the first-order approximation of $E_{jm}^{(j+3)}$ is adopted, i.e.

$$E_{jm}^{(j+3)} = (j+3)\,E_{jm} \;\; , \qquad (2)$$

the relationship (1) becomes also linear for the topography coefficients E_{jm}. Unfortunately, due to large irregularities of the Phobos figure, there are large differencies between the first and higher order approximations (Martinec et al., 1989) and linearized approximation can be used only for evaluating the potential coefficients $\bar{C}_{20}$, $\bar{C}_{32}$, $\bar{S}_{43}$, and $\bar{S}_{51}$ ($\bar{C}_{jm}$ and $\bar{S}_{jm}$ are real and imaginary parts of A_{jm}) for which the contributions of higher order terms reach only units of percentage of linearized approximation terms. For the other potential coefficients linearized approximation is false, the contributions of higher order terms are large. Table 1 gives the Stokes coefficients of the external gravitational potential of Phobos computed according to the fifth-order approximation for the power topography coefficients. The columns of differencies establish the differencies between the fifth-order and linearized approximation.

These differencies clearly show, that the potential coefficients derived by Duxbury (1989) on the basis of linearized approximation are fairly incorrect except four mentioned terms. It is caused by the fact that linearized approximation of non-linear relationship between the external gravity and the topography does not hold for such an irregularly shaped body as Phobos. The linearized formula used by Duxbury (1989) should be replaced by the third-order approximation for spherical harmonics up to the third degree and order and by the fifth-order approximation for higher degree harmonics.

Table 1   The Stokes coefficients of Phobos

j	m	$\bar{C}_{jm}$	$\bar{S}_{jm}$	Difference %	
1	0	0.00543		-98.8	
1	1	-0.00159	0.01473	-165.4	-18.0
2	0	-0.04875		-6.1	
2	1	0.00138	0.00161	-198.5	73.5
2	2	0.02325	-0.00027	15.3	-19.3
3	0	0.00201		-80.4	
3	1	-0.00420	0.00050	-29.0	-244.0
3	2	-0.00851	-0.00063	0.8	-29.5
3	3	0.00220	-0.01318	19.3	24.3
4	0	0.00780		59.5	
4	1	0.00349	-0.00080	-17.1	161.6
4	2	-0.00314	-0.00147	94.2	-25.1
4	3	-0.00280	0.00263	14.3	6.4
4	4	-0.00013	-0.00034	-838.4	49.6
5	0	0.00192		-78.4	
5	1	0.00061	-0.00047	260.9	1.6
5	2	0.00241	0.00013	74.5	-9.3
5	3	-0.00035	0.00298	144.5	84.3
5	4	-0.00197	-0.00048	63.1	19.0
5	5	-0.00114	-0.00187	10.1	103.4
6	0	-0.00071		273.1	
6	1	-0.00199	0.00071	91.0	76.4
6	2	-0.00011	0.00049	-807.7	79.4
6	3	0.00166	-0.00128	72.1	52.9
6	4	0.00016	0.00045	282.0	18.7
6	5	-0.00037	0.00129	136.0	92.6
6	6	-0.00369	-0.00049	60.1	100.1

The gravitational potential model constructed can be used, after including the tidal and centrifugal parts, for any dynamic studies of the Mars-Phobos system.

LONG TERM VARIATIONS IN THE GRAVITY FIELD OF PHOBOS

Because of tidal friction, the mean motion n of Phobos increases. The most recent value of dn/dt determined by Sinclair (1989), is

$$\frac{1}{2} \frac{dn}{dt} = (14.8^{\circ} \pm 2.1^{\circ}) \; cy^{-2} =$$

$$= (2.59 \pm 0.37) \; 10^{-20} \; rad \; s^{-2} \; . \tag{3}$$

The corresponding decrease in the semi-major axis a of Phobos comes out as

$$\frac{da}{dt} = -\frac{2}{3} \frac{a^{5/2}}{(GM)^{1/2}} \frac{dn}{dt} = -(14.22 \pm 2.0) \; 10^{-10} \; m \; s^{-1} =$$

$$= -(4.48 \pm 0.64) \; m \; cy^{-1} \; ; \tag{4}$$

$a = 9\ 378\ 500$ m (present value), $GM = 42\ 828.3 \times 10^9 \; m^3 \; s^{-2}$ is the areocentric gravitational constant. From (2) the product of the Love number k_2 of Mars and the tidal phase lag angle can be obtained:

$$k_2 \varepsilon = -\frac{1}{6} \frac{\left[G(M + m) \right]^{1/2}}{Gm} \left(1 + \frac{Gm}{GM} \right)^{-1} \left(\frac{a}{a'_0} \right)^5 a^{1/2} \frac{da}{dt} =$$

$$= (9.1 \pm 1.4) \; 10^{-4} \; rad = 0.052^{\circ} \pm 0.004^{\circ} \; ; \tag{5}$$

$Gm = 8.4 \times 10^5 \; m^3 \; s^{-2}$ is the phobocentric gravitational constant by Bills and Synnott (1987), $a'_0 = 3\ 397$ km is the mean equatorial radius of Mars. With $k_2 = 0.08$ (Burns, 1972), the phase lag angle is about 0.65° and the estimate for the specific dissipation factor of Mars $Q = 44$. The long term decrease in the energy integral H of the Mars--Phobos system is

$$\frac{dH}{dt} = \frac{1}{2} \frac{GMm}{a^2} \frac{da}{dt} = -(4.4 \pm 0.6) \times 10^6 \; W \; ; \tag{6}$$

the dissipation energy because of tidal friction is of the same order in magnitude (3.8×10^6 W).

The decrease of the Phobos-Mars distance (4) gives rise to the long term variations in the tidal potential V_t due to Mars

$$V_t = \frac{GM}{a} \left(\frac{\rho}{a}\right)^2 \sum_{k=0}^{2} \frac{(2 - \delta_{ko})(2 - k)!}{(2 + k)!} \times$$

$$\times\ P_2^{(k)}(\sin\phi)\ P_2^{(k)}(\sin\phi_o)\ \cos k(\Lambda - \Lambda_o) \tag{7}$$

and in the potential of centrifugal force Q

$$Q = \frac{1}{2}\,\omega^2 \rho^2 \cos^2\phi = \frac{1}{2}\,n^2 \rho^2 \cos^2\phi =$$

$$= \frac{1}{3}\frac{GM}{a}\left(\frac{\rho}{a}\right)^2 \left[1 - P_2^{(o)}(\sin\phi)\right]\ ; \tag{8}$$

(ρ, ϕ, Λ) are phobocentric spherical coordinates (radius-vector, latitude and longitude, respectively) of the potential point (P) on the topography surface of Phobos, (ϕ_o, Λ_o) phobocentric latitude and longitude of the mass center of Mars (O'), $P_2^{(k)}(\sin\phi)$ is the associated Legendre function of the second degree and k-th order, δ_{ko} is the Kronecker delta, ω is the angular velocity of Phobos' rotation (because of synchronous, $\omega = n$). Because the inclination of the orbital plane of Phobos to the Martian equator is small ($i = 1.02^o$), we put $\phi_o = 0$; let the prime meridian plane of Phobos pass through O', then $\Lambda_o = 0$. The long term variation in $(V_t + Q)$ at P reads

$$\frac{d(V_t + Q)}{dt} = -\frac{GM}{a^2}\left(\frac{\rho}{a}\right)^2 \left[1 - \frac{5}{2}P_2^{(o)}(\sin\phi) + \right.$$

$$\left. + \frac{3}{4}P_2^{(2)}(\sin\phi)\ \cos 2\Lambda\right]\frac{da}{dt}\ . \tag{9}$$

No variations occur if

$$\frac{5}{2}P_2^{(o)}(\sin\phi) - \frac{3}{4}P_2^{(2)}(\sin\phi)\ \cos 2\Lambda = 1\ , \tag{10}$$

e.g., if

$$\cos 2\Lambda = 1, \qquad |\phi| \gtrsim 60^0,$$
$$\cos 2\Lambda = 0, \qquad |\phi| \gtrsim 50.77^0,$$
$$\cos 2\Lambda = -1, \qquad |\phi| \gtrsim 0^0.$$

The corresponding increase in the radial derivative of (9), i.e. decrease in the total gravity at P is

$$\frac{dg}{dt} = -2 \frac{GM}{a^3} \frac{\varrho}{a} \left[1 - \frac{5}{2} P_2^{(0)}(\sin\phi) + \frac{3}{4} P_2^{(2)}(\sin\phi) \cos 2\Lambda \right] \frac{da}{dt} ; \qquad (11)$$

numerically, for P situated on the equator and prime meridian (ϱ = 13 220 m, ϕ = 0^0, Λ = 0^0)

$$\frac{dg}{dt} = -3 \times 10^{-9} \text{ m s}^{-2}/\text{cy}, \quad \frac{1}{g}\frac{dg}{dt} = -1.3 \times 10^{-6}/\text{cy}. \qquad (12)$$

The decrease in g is relatively large. The zero-gravity at P (ϱ = 13 220 m, ϕ = 0^0, Λ = 0^0) should start at $a(\Delta t) \doteq$ $\doteq$ 6 550 km, in time Δt. Integration of (5) gives the estimate for Δt as

$$\Delta t = \frac{2}{39} \frac{(GM)^{1/2}}{Gm} \frac{Q}{k_2} \frac{\left[a(\Delta t)\right]^{13/2} - a^{3/2}}{{a'_0}^5} \doteq 3 \times 10^7 \text{ y}.$$

CONCLUSIONS

1. Even if homogeneous body, the relationship between topography and gravity potential harmonics of Phobos is not linear.

2. Using the most recent value of acceleration of Phobos by Sinclair (1989), the long term decrease of the equatorial gravity at the prime meridian is about 10^{-6}/cy (in gravity of Phobos units).

3. The zero gravity at the point in question will occur in 30 million years and the disintegrating process of the body of Phobos will start at the equatorial zone.

REFERENCES

Bills, B. G. and Synnott, S. P. (1987). Planetary Geodesy.
    Rev. Geophys. 25, 833.
Burns, J. A. (1972). Dynamical characteristics of Phobos
    and Deimos. Rev. Geophys. 10, 463.
Burša, M. and Martinec, Z. and Pěč, K. (1988). Principal
    moments of inertia, secular Love number and origin of
    Phobos. Presented at XXVII COSPAR - Espoo, Finland 1988.
Duxbury, T. C. (1989). The figure of Phobos. Icarus 77,
    (in print).
Duxbury, T. C. and Callahan, J. D. (1989). Phobos and Dei-
    mos control networks. Icarus 77, 275.
Martinec, Z. and Pěč, K. and Burša, M. (1989). The Phobos
    gravitational field modelled on the base of its topo-
    graphy. Earth, Moon and Planets, (in print).
Sagitov, M. U. and Tadzhidinov, Kh. T. and Mikhajlov, B.O.
    (1981). Model gravitacionnogo polja Fobosa. Astron.
    vest. XV, 142.
Sinclair, A. T. and Jones, D. H. P. and Williams, I. P.
    (1989). Secular acceleration of Phobos confirmed from
    positions obtained on La Palma. Mon. Not. R. astr. Soc.
    237, 15.
Turner, R. J. (1978). A model of Phobos. Icarus 33, 116.

"FELDGRAV"
A DATA BANK SYSTEM FOR HIGH PRECISION GRAVITY OBSERVATIONS

Ernö Czuczor
Hessisches Landesvermessungsamt Wiesbaden, FRG
Carl Gerstenecker
Technische Hochschule, Darmstadt, FRG

Abstract: Since 1987 the relational data bank system "FELDGRAV" is in use by the State Geodetic Survey of Hessen, FRG. The data bank system is written in TURBO-PASCAL. A data logging system allows automatic data collecting within the data bank system. The software runs on IBM compatible portable Personal computers (PC). The ad hoc status of the system is described, whereby special emphasis is given to the adjustment procedure. Advantages of the system are seen in considerable acceleration of the observations and computations.
Achieved accuracy improvement within a high precision gravimetric network was about 10%.

INTRODUCTION

Progress in modern electronics makes a lot of different automatic data logging systems available. The application of such data logging systems (DLS) for relative gravity observations in the field is not new (Gerstenecker, 1988). The goal is to improve speed and accuracy of the measurements and to reduce the frequency of errors.
A data logging system on the basis of a battery operated portable lap top computer was developed and tested in the field since 1987 by the State Geodetic Survey of Hessen. Components of the system and its configuration are described in (Czuczor et al., 1989).
Experiences with the system have shown, that the hard ware was working properly well in the field. However the system was inflexible concerning data handling (preparation, correction, adjustment). On the other side the system didn't fully exploit all options offered by the PC. The idea cames up, to integrate the data logging system within a relational data bank system running on the PC.
The data bank system was named "FELDGRAV". Programming language is TURBO-PASCAL under the operating system MS-DOS 2.11 and higher.
The data basis of the data bank system (Fig.1) consists of four relations
- GRAVITY_POINT
- OBSERVATION
- INSTRUMENT
- MODUL.
Three applicationprograms
- DATA_ADMINISTRATION
- DATA_COLLECTION
- DATA_ADJUSTMENT

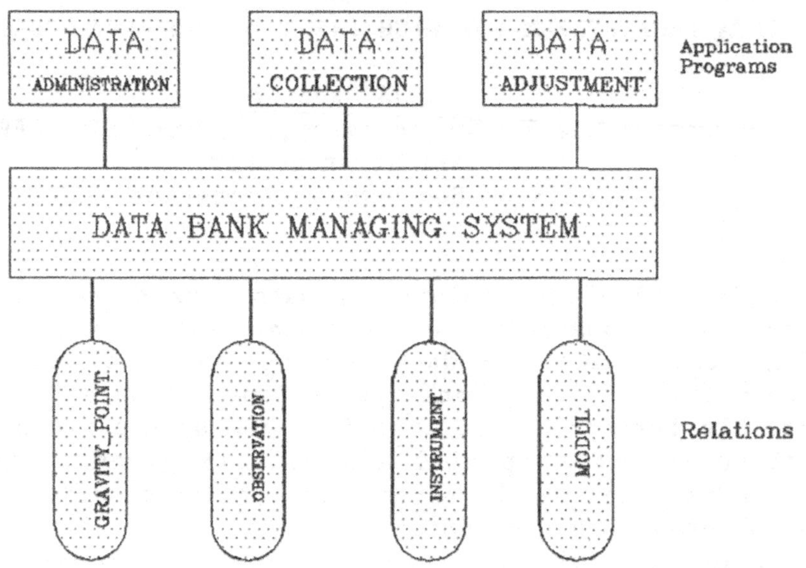

Figure 1: Data Bank System "FELDGRAV"

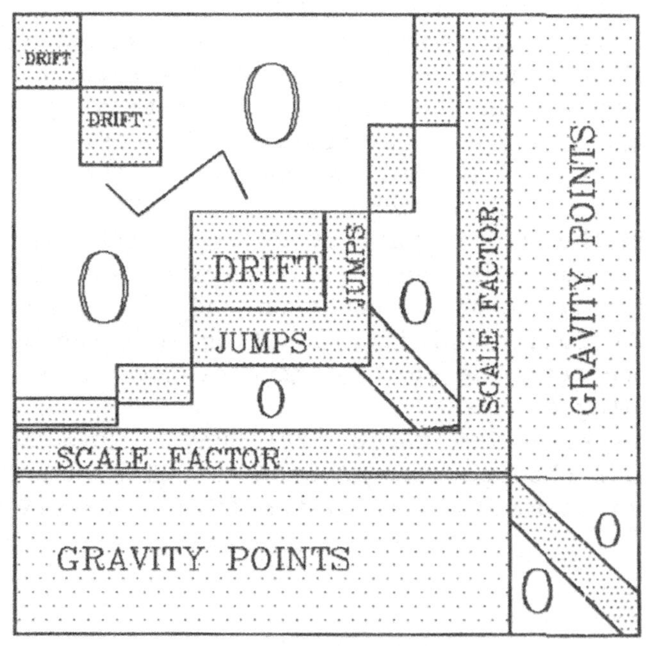

Figure 2: Structure of N_Matrix

have access to the relations via the data bank managing system.
Attributes of the relations and options of the programs
DATA_ADMINISTRATION and DATA_COLLECTION are described in (Czuczor et
al.,1989) as well as physical dimensions, hard- and software limits and
disadvantages of the data bank system. In the following section we will
give special features of the program DATA_ADJUSTMENT. Experiences concerning accuracy and speed of the system gained during 2 field campaigns
in Turkey and laboratory tests will complete this paper.

DATA ADJUSTMENT IN "FELDGRAV"

Least square estimation of the unknown parameters is carried out in the
Gauss Markoff model. The observation equation is written as

$$v_{i,j,k} = G_j + \sum^{p} S_{k,p} * Z^{p}_{i,j,k} + N_{k,1} + \sum^{m} a_{k,m} * t^{m}_{i} - A_{i,j,k} \qquad (1)$$

with

$v_{i,j,k}$ = residual of observation A_i at time t_i, station j with
gravimeter k

G_j = unknown gravity value of station j

$S_{k,p}$ = unknown scale factor of degree p

$Z^{p}_{i,j,k}$ = reading of gravimeter k at time i and station j in counter
units

$N_{k,1}$ = 1 th unknown constant drift term (jump) of gravimeter k

$a_{k,m,r}$ = r th unknown drift coefficient of gravimeter k and degree m

t_i = time of reading $A_{i,j,k}$.

$A_{i,j,k}$ = reading of gravimeter k at station j and time i in $1*10^{-5}$
m sec^{-2}

Free network adjustments as well as adjustments with constraints are
possible. Because of the RAM-limitation of TURBO-PASCAL to 64 KBytes for
adjustment problems with more as 140 unknowns sparse matrix techniques
must be applied.

Obviously the matrix of the normal equation N can be ordered in a tree
structure ("Helmert blocking") as shown in Figure 2. Grepel(1987) has
collected efficient storage and inversion algorithms for sparse
matrices. With such techniques the solution of equation systems with up
to 500 unknowns is possible within "FELDGRAV" on a lap top PC as the
Toshiba T1000. The number of observations is only limited by the available external storage capacity of RAM-, hard- and floppy disks. Computing
time is mainly dependend on the access speed of the external storage
units. For the least square adjustment of about 789 gravity observations
and 143 unknown parameters the Toshiba T1000 needs about 900 seconds.

LABORATORY AND FIELD TESTS

An important motivation developing "FELDGRAV" was the request on impro-
ved accuracy of gravity observations.
Automatic data logging can improve the reading accuracy of observations.
Informations, however about the accuracy of the data logging system

(DLS) in "FELDGRAV" in comparison to conventional observation methods are hardly to obtain. Relative gravity measurements are more disturbed by environmental effects and unresolved instrumental problems as by inaccurate gravimeter readings. Improving the reading accuracy doesn't diminish the mean square error of the unit weight of a least square adjustment significantly.

Investigating reading accuracy special experiments were carried out in the laboratory as well as in the field, to find the best read out procedure for an LaCoste-Romberg model G gravimeter. 5 different read out procedures were applied:
- optical reading
- electronic reading with external digital voltmeter
- electronic reading with DLS and averaging over 5 seconds
- electronic reading with DLS and moving averaging over 120 seconds
  (window length 30 sec)
- electronic reading with DLS and digital filtering over 180 seconds
  (Butterworth lowpass filter of degree 4, cut off frequency 0.025 Hz)

For the experiments the gravimeter LCR-G 563 was used. The gravimeter was unpacked, levelled, unclamped, readed, clamped and packed again 30 times within 1 observation series. Results are given in Table 1.

Table 1 : Accuracy of different read out procedures

| Read out method | Optical | External voltmeter | DLS | | |
			Average	Moving average	Digital filter
Period of 1 observat.	30 sec	30 sec	5 sec	120 sec	180 sec
Standard deviation Laboratory:	50	40	35	< 10	< 10
Field: $[nm*sec^{-2}]$	55	40	40	30	30

The high reading accuracy in the laboratory of the DLS with moving averaging or digital filtering procedures is lost in the field probably due to temperature changes and environmental effects as sunshine or wind and the long necessary observation period. We therefor favour the simplest and quikest averaging read out method.

In a local gravimetric network in Turkey two LaCoste-Romberg gravimeter were read with DLS. One instrument was observed with the electronic output system of the gravimeter and an external voltmeter, one instrument was read using the optical microscope. All instruments were transported in the same car.

Table 2 shows the mean square error of the unit weight of the different instruments.

Table 2: **Mean square errors of unit weight**

Instrument	Mean square error of unit weight nm* sec^{-2}	Observation method
LCR-G 258	82	DLS
LCR-G 563	67	DLS
LCR-G 839	124	external voltmeter
LCR-G 874	134	microscope

From table 2 we can conclude, that significant higher accuracy is obtained with the automatic data logging system (DLS). However, if LCR-G 839 is readed in the same manner as LCR-G 258 and LCR-G 563 we get the same ratio of the mean square errors of the unit weight. It is more fair, to claim for a similar or slightly better accuracy ($\leq$10%) of observations with DLS. Significant improvement of the measuring accuracy is not achieved by improving reading accuracy of gravimeters.
The acceleration of the observation procedure depends on the chosen read out method. If simple averaging is used, the observation time is shortened about $\leq$ 10%.
The biggest advantage of "FELDGRAV", however is the acceleration of data analysis. Final results of field campaigns extending over several weeks are nearly "on line" available as well as the results of measuring spindulum investigations. Conventional manual data processing cannot compete with "FELDGRAV".

CONCLUSIONS

Even if the development of the data bank system "FELDGRAV" is still not finished, its advantages are obvious:
- high reading accuracy
- slight improvement of measurement accuracy
- considerable acceleration of adjustment procedures.

REFERENCES

Czuczor,E., Falk,H., C.Gerstenecker (1989): Automation in der
    Feldgravimetrie, Zeitschrift für Vermessungswesen, Vol. 114, pp 259-
    267
Gerstenecker, C. (1988): An Intelligent Data Collecting System for
    Relative Gravity Measurements, Bulletin D'Information, No.62,pp 51-
    56, Bureau Gravimetrique International, Toulouse
Grepel, U. (1987): Effiziente Rechenverfahren für umfangreiche

geodätische Parameterschätzungen, Mitteilungen aus den Geodätischen
Instituten der Rheinischen Friedrich-Wilhelms-Universität, Bonn

DIGITAL COMPILATION OF GRAVITY DATA OVER THE JAPANESE ISLANDS

Yoshiteru Kono and Nobuhiro Furuse
Department of Earth Sciences, Faculty of Science, Kanazawa
University, Kanazawa, 920, Japan

INTRODUCTION

The Japanese Islands are situated in the most tectonically active areas of the Pacific and Philippine Sea subduction zones. The high mountain regions have many summits of 3,000 meters in elevation, comparable with the Alps. The length of the islands is about 2,000 km. The topographic relief is so great that roads are largely restricted to river courses, and large areas around high mountains remain difficult to access for gravity measurements.

Gravity measurements in Japan were started in the last century. In 1899, Dr. H. Nagaoka connected the absolute gravity in Potsdam with Tokyo by means of a Strerneck pendulum. By 1915, more than 100 gravity stations were established on the Japanese Islands.

The first gravity measurements in an oceanic area was carried out near the Japan trench axis by Dr. M. Matsuyama in 1932. Fig.1 shows his preliminary results (Matsuyama, 1934). A submarine and a Vening Meinesz type pendulum No.4 were employed for the measurements.

After the second world war, gravity measurements by means of spring-type gravimeters increased rapidly. In 1954, gravity measurements at first and the second order leveling stations along main national roads were performed by Dr. C. Tsuboi and his colleagues (Tsuboi, 1954) using a Worden gravimeter to measure 1,500 points.

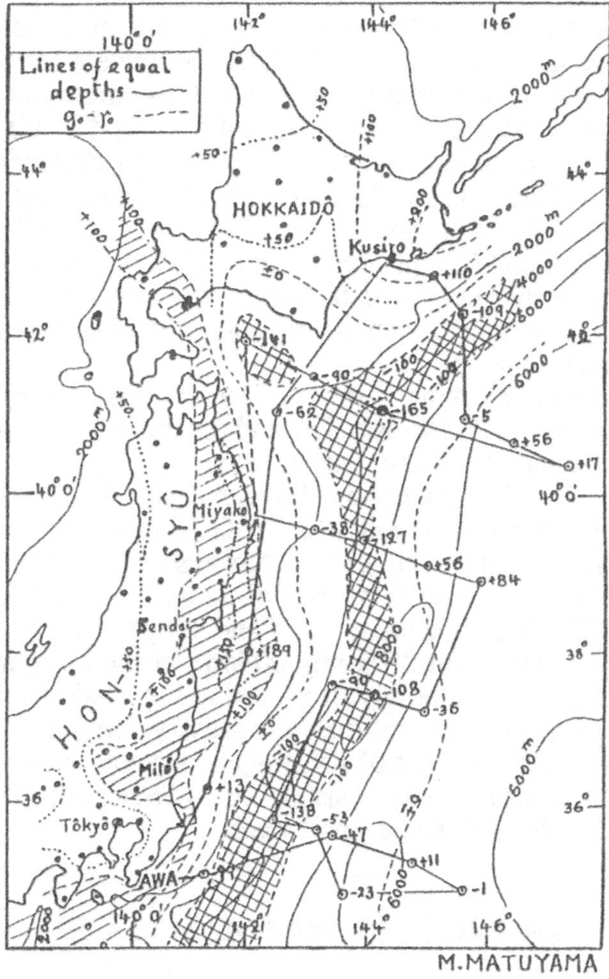

Fig.1 Preliminary results of gravity measurements near the Japan trench conducted by Dr. M. Matsuyama (Matsuyama, 1934).

In 1969, the Geographical Survey Institute completed precise gravity measurements one round all over Japan. However, gravity stations were still restricted to main national roads, and wide vacant areas remained everywhere else.

Gravity measurements over oceanic areas were started by Dr. Y. Tomoda (Tomoda, 1972; Tomoda and Fujimoto, 1982) with a newly developed precise surface ship gravimeter.

The Hydrographic Department of Maritime Safety Agency and the Geological Survey of Japan also were making marine gravity measurements, particularly over the continental shelf around Japan.

In parallel with these nation-wide gravity measurements, detailed local measurements of gravity for geophysical exploration were also carried out in various potentially economic areas.

GRAVITY DATA SOURCES

There are four groups of data source:

The first is the data measured by ourselves during the last 15 years, numbering more than 10,000.

The second is the data measured by the Geographical Survey Institute (GSI) and by the Hydrographic Department for geodetic purposes. These data cover all areas in and around the Japanese Islands. The GSI maintains a gravimetric reference frame throughout Japan since the Institute is the national organization responsible for geodetic gravity data in Japan. There are about 130 first order and 15,000 second order gravity stations.

The third is data measured by university researchers for various academic purposes.

The fourth is the data measured by govermental exploration organizations such as the Geological Survey, the New Energy and New Technology Development Organization, the Metal and Mining Agency, etc.. However, all data from private exploration companies and large numbers of data from some govermental organizations are not available for the present compilation. Most data in this category are unpublished. Each data set covers local areas in detail.

Gravity systems, formats of records, and descriptions of related information are different for each data source. The authors unified them as if they were measured and processed by one institution as described later.

Most data were obtained by means of Worden, North American and LaCoste & Romberg gravimeters on land, and TSSG (Tokyo Surface Ship Gravimeter), LaCoste & Romberg, and Bodensee ship gravimeters at sea. The oldest data set is 1951.

The total number of data digitally compiled is about 500,000 points of which about 100,000 points are on land and 400,000 are at sea. The average density of measurement points is about 25 point/100 km2, or the mean spacing between points is about 2 km. All measurement points are plotted in Fig.2. where each dot represents a measurement point.

Higher concentrations of gravity measurements are still required on land for the southeastern part of Kyushu,and several areas in Honshu. Required areas more urgently are marine gravity measurements in Setonaikai (Seto inland sea), in an area off northeastern Honshu, and some smaller areas.

DATA PROCESSING AND DATA EXAMINATION

Each data set is compared with the GSI's data set and both biases of gravity values and inaccuracy of gravimeter coefficients are examined. During these processes, incorrect or suspected data are either corrected or rejected from the data set. After these examination, all data are adjusted to the International Standardization Net 1971 (IGSN71) system.

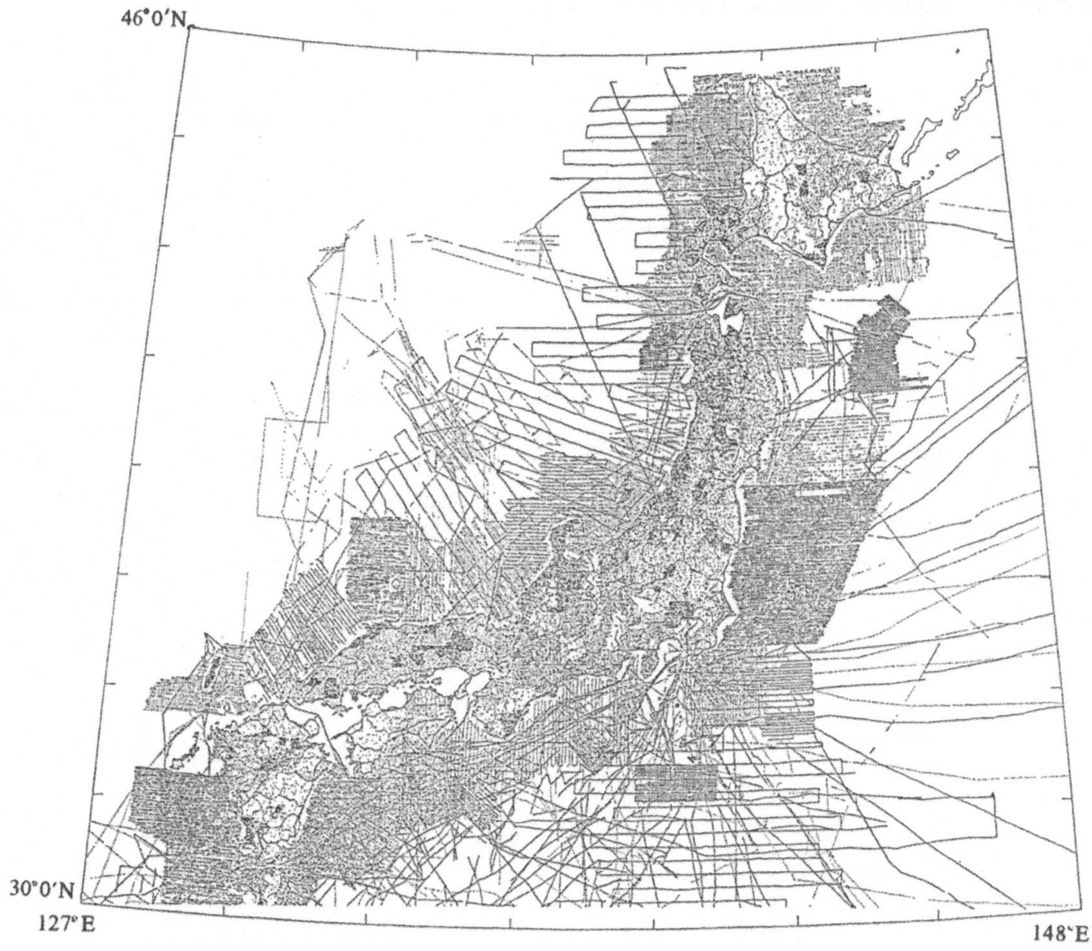

Fig.2 Distribution of individual gravity measurements in and around the Japanese Islands.

On land, gravity anomalies are expressed by Bouguer anomalies, Δg", and are calculated as follows

$$\Delta g" = g - \gamma + Atm + 0.3086h - 2\pi G\rho h + Tr \qquad (1)$$

where g, the gravity value, γ, the normal gravity from the geodetic reference system 1967 as expressed in (2), Atm, atmospheric (mass) correction term as expressed in (3), G, the universal gravitational constant, h, the elevation of an observed point (units of meters above sea level) and Tr, the terrain correction term.

Gravity and gravity anomalies are expressed in milligal (mgal) in this study (1 mgal = 0.1 G.U. = 10^{-5} m.s^{-2}),and

$$\gamma = 978031.85(1 + 0.005278895\sin^2\varphi + 0.000023462\sin^4\varphi) \qquad (2)$$

$$Atm = 0.87 - 0.965 \times 10^{-3}h \qquad (3)$$

The density for calculation of Bouguer anomaly is postulated to be 2.67 g.cm^{-3}, as usually employed in geodesy. Measured rock densities for pre-Tertiary sedimentary rocks, volcanic and metamorphic rocks in the Japanese Islands vary from 2.6 to 2.7 g.cm^{-3} on average.

Fig.3 An example of the gravity anomaly map in and around the Japanese Islands. Original scale is 1:1,000,000. Contour interval is 2 mgal.

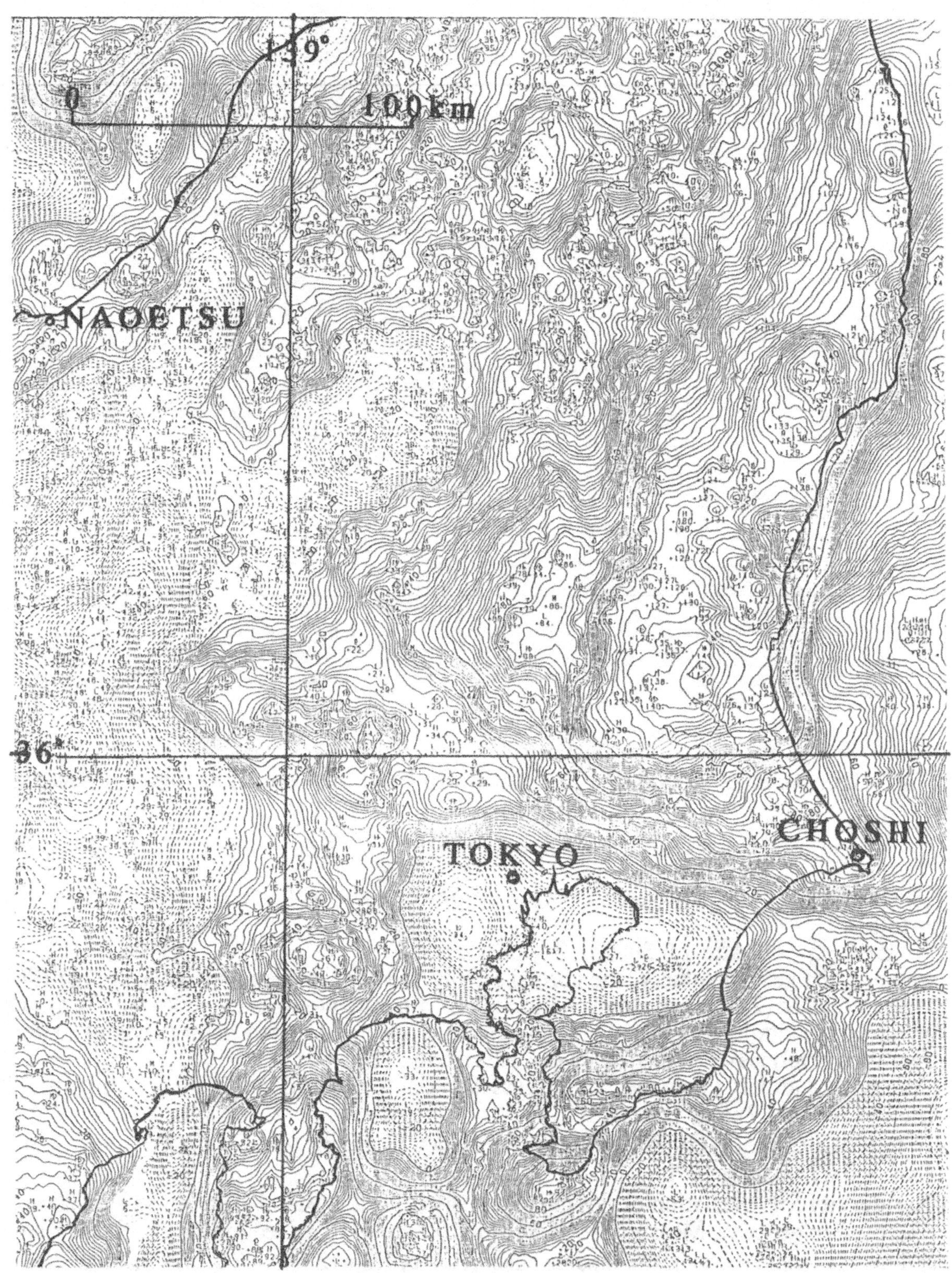

On land, terrain correction terms are calculated by using digital topographic data on a grid with a spacing of 20"(NS)x22.5"(EW) (about 500m x 500m) and the accuracy of height is within 10m. More than 0.3 million points within 40km of each observed point are used for the calculation (Kono and Kubo, 1983). The accuracy of terrain correction terms is 10 % or less.

At sea, the gravity anomaly is expressed by the free-air anomaly ($\Delta g'$):

$$\Delta g' = g - \gamma + Atm + 0.3086h \qquad (4)$$

For h=0, (4) gives:

$$\Delta g' = g - \gamma + 0.87 \qquad (4')$$

Calculated gravity anomalies are converted into grid data at intervals of 1'(NS) x 1'(EW) (about 1.8km x 1.5 km). A contoured anomaly map with 2 mgal interval is generated by using an automatic contouring technique.

A part of the gravity anomaly map thus generated is illustrated in Fig.5. The figure covers the central part of Honshu island, with Tokyo in the centre.The map is prepared at a scale of 1:1,000,000 with a contour interval of 2 mgal. Coloured (25) full size maps with an explanetary text will be published soon (Kono and Furuse, 1989).

GRAVITY EFFECT DUE TO SUBDUCTING PLATES UNDERNEATH THE JAPANESE ISLANDS

The subducting Pacific and Philippine Sea plates underneath the Japanese Islands (slabs) are estimated to be denser than the surrounding mantle (asthenosphere) by about 0.065 gr.cm^{-3}. This adds to the observed gravity at the earth's surface. Fig.4 explains the relationship between the subducting plates, the gravity contribution due to slabs, the observed gravity anomaly, and the residual gravity anomaly.

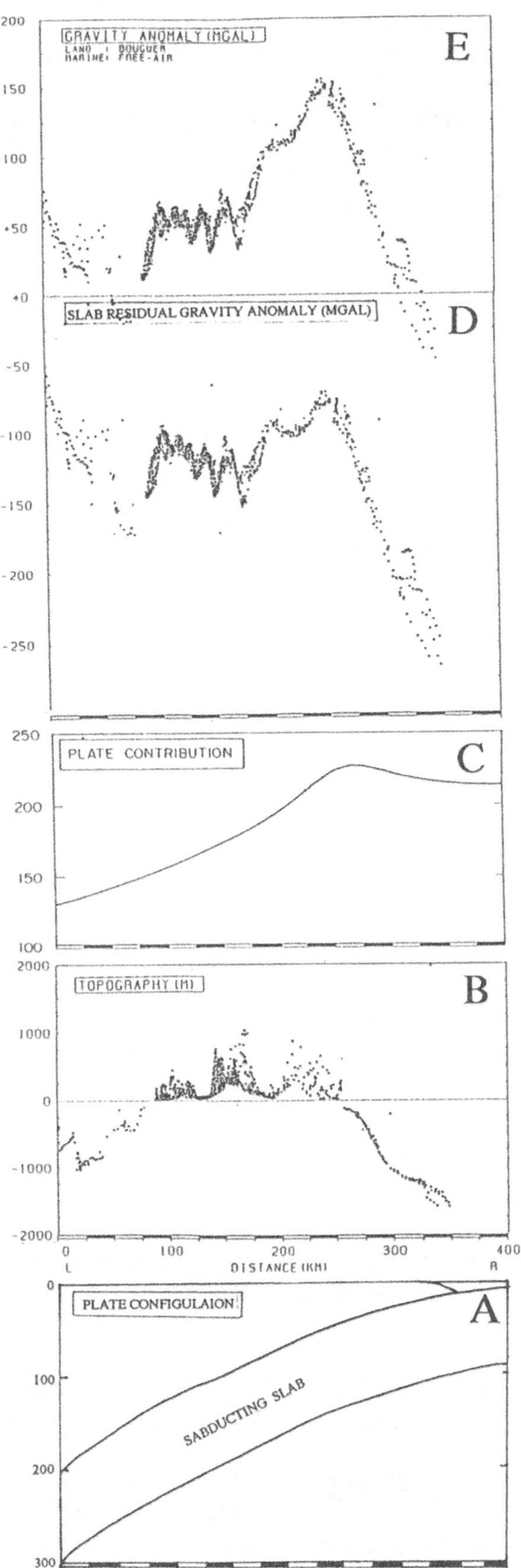

Fig.4 Subducting plate (A),topography (B), gravity contribution due to the slab (c) , residual gravity anomalies (D) and observed gravity anomalies (E).

The authors calculated the gravity contribution due to the subducting plates, taking into
account a 3–dimentional configuration of the plates and an effect of earth's curvature
(Furuse and Kono, in preparation). According to the calculations, the gravity contribution
reaches 220 mgals near the Pacific coast on the northeastern part of Japan and decreases
towards the Japan Sea (Fig.5). If we subtract this from the observed gravity anomalies, the
reduced gravity anomalies (we call this "Slab Residual Gravity Anomaly: SRGA) should be
mainly due to crustal structures. An example of cross sections of observed gravity
anomalies and topography and Slab Residual Gravity Anomalies is shown in Fig.6.

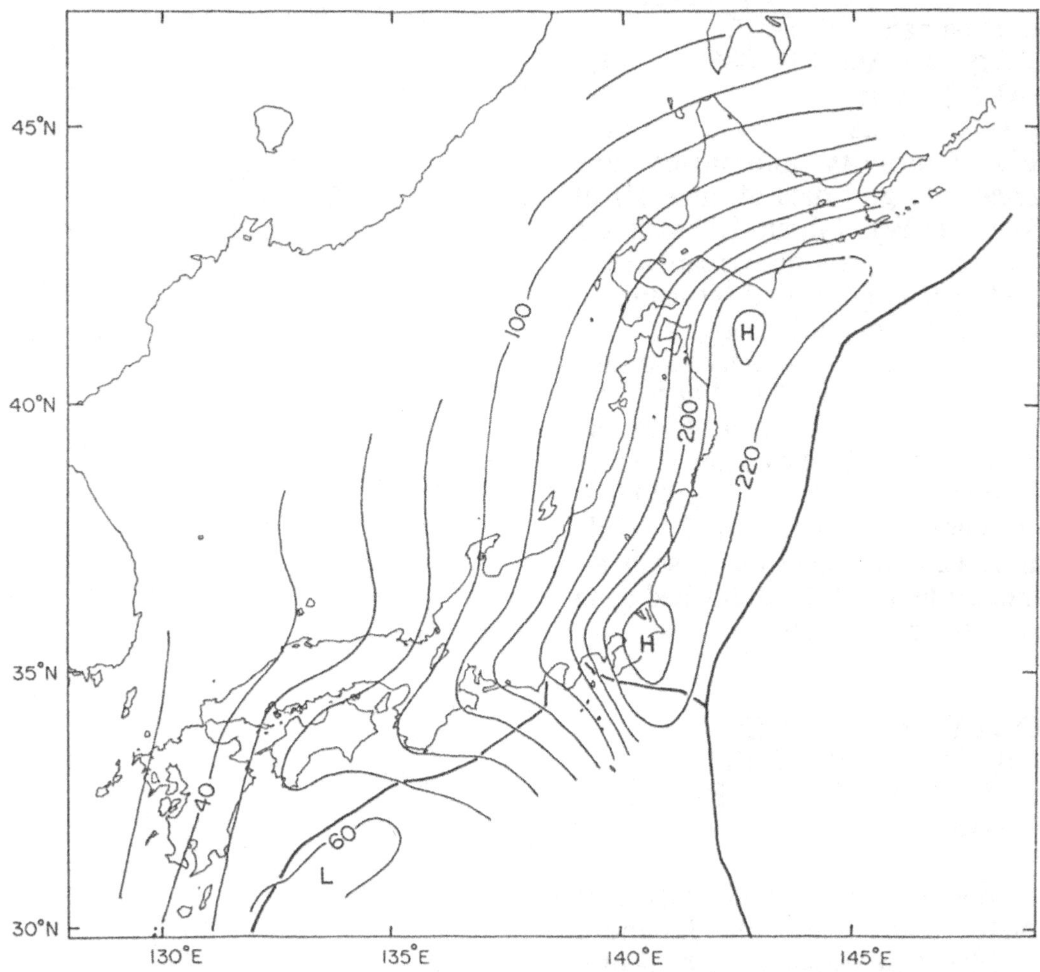

Fig.5 Theoretical distribution of gravity contributions due to subducting the Pacific and
Philippine Plates (slabs).

SOME TYPICAL CHARACTERISTIC FEATURES OF GRAVITY ANOMALIES OVER THE JAPANESE ISLANDS

The highest gravity anomaly on land, 220 mgal, appears around the Nemuro peninsula,
Hokkaido island. The maximum in Honshu island is 180 mgal on the northeastern part of
the island, the Sanriku coast. The minimum Bouguer anomaly of –110 mgal appears west
of Hidaka mountains, Hokkaido. The minimum value in Honshu, –75 mgal, is observed
around the Hida mountains, central Honshu. The minimum free-air anomaly of –300 mgal

appears at the junction of the Japan and the Izu–Mariana trenches. Therefore, the variation of gravity anomalies around the Japanese islands is about 500 mgal. In the central part of Honshu , the highest (about 200 mgal) and the lowest (–300 mgal) free-air anomalies are only 160 km apart.This may be one of the largest gravity gradients in the world.

The most remarkable change in the distribution pattern on land is recognized around the central part of Honshu, a zone between the towns of Naoetsu (Japan Sea side) and Chosi (Pacific Ocean side). The authors named this the "Naoetsu-Choshi Line" (Na o e tsu - Chou shi Line in pronounciation) from gravity anomaly point of view. Another impressive change of pattern is observed or northern Honshu. The elongate pattern in the north-south direction divides the area into the Pacific and Japan Sea sides. The boundary was named the "Morioka-Shirakawa Line " by Dr. C. Tsuboi in 1954. The geological implications of these foundaries will be discussed elsewhere.

Regional negative Bouguer anomalies on land are observed on the central part of Hokkaido, the central part of Honshu, and the adjacent parts of western part of Honshu, northwestern Shikoku, and Kyushu islands, respectively. Negative anomalies in central Honshu consists of two parts: one related to the Hida mountains, the other, to Lake Biwa. Among them, only one negative anomaly, the Hida mountains area may be interpreted using the concept of isostasy, because the other areas have no high mountain ranges related to regional negative anomalies. In another words, there is no relevant correlation between negative Bouguer anomalies and topographic relief. Even if we reduce the gravity contribution due to subducting plates from observed anomalies, the relationship as mentioned above does not change. These facts may indicate the tectonic states of the Japanese Islands.

More detailed studies will be required after the reduction of the gravity contribution due to subducting plates.

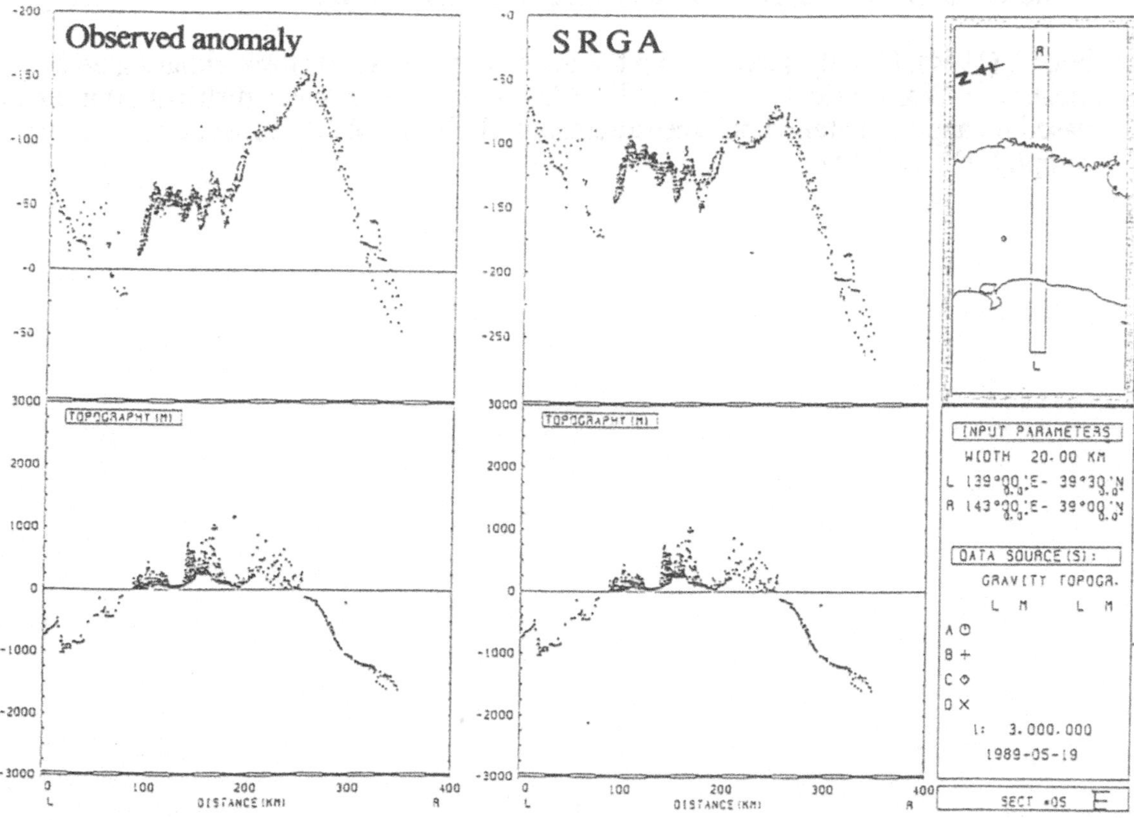

Fig.6 An example of profiles of observed gravity anomalies and topography and slab residual gravity anomalies in the same section.

Acknowledgment.

Dr. T. Lewis kindly read the manuscript and improved it. Dr. T. Nagao and Miss M. Kono helped us for the preparation of the manuscript. we acknowledge them.

REFERENCES

Kono, Y. and Kubo, M.(1983). Calculation of terrain correction term by using meshed mean height data, *J. Geod. Soc. Japan,* 29, 101-112. (in Japanese).

Furuse, N. and Kono,Y. Gravity contribution due to descending lithospheres beneath the Japanese Islands. (in preparation).

Kono, Y. and Furuse, N. (1989). Detailed gravity anomaly map in and around the Japanese Islands with transparent seismicity maps—Scale 1:1,000,000 (tentative title), University of Tokyo Press. (in press).

Matsuyama, M. (1934). Measurements of gravity over the Nippon Trench on boad the I. J. Submarine Ro-57, *Proc. Imp. Acad. Japan,* 10, 626-628.

Tomoda, Y. (1972). Free air and bougurer gravity anomalies in and around Japan, University of Tokyo Press.

Tomoda, Y. and H, Fujimoto. (1982). Maps of gravity anomalies and bottom topography in the Western Paciffic, Bull. Ocean Res. Inst., Univ. Tokyo.

Tsuboi, C. (1954). Gravity survey along the lines of precise levels throughout Japan by means of a Worden Gravimeter, part IV. Map of bouguer anomaly distribution in Japan based on approximately 4,500 measurements. Bull. Earthq. Res. Inst., Univ. Tokyo, Suppl.4, III, 125-127

Gravity Measurement on the Bottom of Toyama Trough

Takeshi Matsumoto and Hiroshi Hotta
Japan Marine Science and Technology Center
Yokosuka, 237, Japan

INTRODUCTION

Recently, the eastern margin of the Japan Sea is remarked as a newly developed convergent boundary between Eurasian and North American plates (Nakamura, 1983). The assumed boundary is located from Fossa Magna along the Toyama Trough, Sado Ridge, Okushiri Ridge towards the western egde of Okhotsk Sea (Fig.1). Some features of subducting slab, that is, thrust faulting, are observed by seismic profiling survey in several places along the line, (Nakamura, 1983; Asada et al., 1989). The southernmost part of the Toyama Trough, however, shows no evidence of thrust faulting from the result of seismic profiling, although this is located just north of Fossa Magna, and so some active tectonic movement is expected. In fact, this place is characterized as low seismicity area on the whole, but recently plenty of earthquakes of M3 $\sim$ 5 occur frequently, and the largest one (M5.9) was occured on March 24, 1987. The distribution of the hypocentres suggests the existence of thrust faulting across the Toyama Trough off Noto Peninsula according to the seismic data catalogue of Japan Meteorological Agency. Therefore, this area of the expected plate boundary was selected as a target of precise survey, and gravity measurement was conducted on the sea bottom across the trough by use of the Japanese deep sea research vessel "SHINKAI 2000" in 1987 and 1988. The authors would like to report the result of the sea bottom gravimetry at the Toyama Trough and discuss the sub-bottom structure derived from the gravity anomaly and its tectonic implications.

OUTLINE OF GEOLOGICAL AND MORPHOLOGICAL STRUC-

TURE OBTAINED BY PRE SITE SURVEY AND VISUAL OBSERVATION FROM THE SUBMERSIBLE

Before the precise survey by the submersible, a pre site survey cruise was carried out by R/V KAIYO in 1986. In this cruise, (1) topographic survey by use of Seabeam, (2) high-precision topographic and bottom surface geological survey by use of JAMSTEC Deeptow side-scan sonar, and (3) visual observation by use of JAMSTEC Deeptow camera were conducted in this area in order to determine the precise survey points by the submersible.

Seabeam map shows large-scale meandering of Toyama Deep Sea Channel running at the western edge of Toyama Trough in this place (Fig.2). This channel is considered to be formed mainly by erosion like rivers on land judging from its feature. However, NNE-SSW trending structure is quite predominant along this deep sea channel unlike normal meandering rivers on land.

The result of the visual observation by the submersible and deeptow camera at the deep sea channel shows that, (1) The bottom of the Toyama Deep Sea Channel located along the western margin of the Toyama Trough is quite flat and is covered with soft sediment. Pebbles are found on the bottom in several places. (2) A layer made up of round pebbles is formed below 1500m in water depth, and semi-consolidated silt bed lies above the pebble layer. (3) The same kinds of beds are observed on both eastern and western slopes of the deep sea channel. (4) The bottom of the central and eastern parts of the Toyama Trough is widely covered with sediments. No traces of crustal movements nor exposition of basement rocks are recognized even on the high gradient slopes shown on the topographic map. Rather steep and gentle slopes of sediment appear alternately in such a place.

SEA BOTTOM GRAVITY MEASUREMENT

Gravity measurement on the sea bottom was conducted along the WNW-ESE line across the Toyama Trough using La Coste G-183 gravimeter. This gravity meter was used in several places of both high and low latitude stations through the project of gravimetric connections along the Circum-Pacific zone, and the scale constants were updated in this project (Nakagawa et al., 1983). The new scale constants were used for processing of raw gravimetric data. Relative measurement based on the Toyama gravity base station (in Toyama Meteorological Observatory) was conducted for sea bottom gravimetry.

The obtained gravity data are to be reduced to the values on the sea surface to compare with those by the surface ship measurement. Free air anomaly reduced to the mean sea level is expressed as

$$\Delta g' = g_{obs} - \beta D + 4 \pi k^2 \rho_w D - \gamma$$

where g_{obs} is observed gravity on the sea bottom, γ is normal gravity at the station, β is gravity gradient (dg/dz), k^2 is gravitational constant, D is water depth of the measured point, ρ_w is density of sea water. The equation, however, is applied in case that gravity effect by the undulation of bottom topography is negligible. When the gravity effect by the bottom topography is taken into account, terrain correction should be applied with the assumption of density of bottom material, which is the sum of the upward gravitational force at the sea bottom station by the part above the level of the station and the downward one by the same part at the sea surface just above the station (Satomura et al., 1987). In this work, two-dimensional terrain correction using the Seabeam bathymetric data around the survey points are conducted. Density for the terrain correction is assumed as 1.8g/cc, as discussed later.

Constant gravity gradient (dg/dz) of 0.3086mgal/m is assumed as for free air reduction of land gravimetry. Using the same value would result in not so large error near the coast. But the value of dg/dz on the deep sea must be quite different, and so comparison between the gravity corrected by this value and that of surface ship measurement would be unavailable due to a large error. However, if the gravimetric mission is restricted at a narrow area and on the deep sea only, and if the difference of the water depth among the measurement points is not large, then the result of sea bottom gravimetry can be compared with that of surface ship gravimetry by use of the relative values, although absolute values are not compared with each other. Water depth is measured by CTD depth sensor equipped on the submersible, so the real value of the water depth at the site is 2.27m deeper than the depth expressed by CTD readout value. Total depth correction including both dg/dz and gravity effect by the sea water above the station is 0.2222 mgal/m.

RESULTS

Fig.2 shows the result of measurement at the Toyama Trough. Dots are stations, and the numbers are free air anomaly reduced to the sea surface. In this work, for convenience, two-dimensional terrain correction across the trough is applied to the observed data. If the correction density is assumed to be 2.67g/cc as usually used, free air anomaly reduced to the sea surface becomes maximum above the axis of the deep sea channel. This correction density is obviously too large. Fig.2 shows the result with the correction density of 1.8g/cc, which minimises the correlation between the bottom topography and corrected free air anomaly at the deep sea channel. Considering that the whole area is covered with soft sediment as shown by JAMSTEC Deeptow camera and the submersible, the value of the correction density is quite adequate and reasonable. The result shows that free air anomaly decreases from the edge towards the center of the trough. It is about -10mgal at the deep sea channel, whereas the central axis of the trough in this part presents about -42mgal.

The obtained gravity data were compared with the surface ship gravity data (Japan Hydrographic Department, 1968) above this track. Fig.3 shows the result. According to the surface ship gravity data, no correlation at the topographic depression of Toyama Deep Sea Channel has been observed, and the minimum of free air anomaly exists at the central part of the Toyama Trough. The result of the sea bottom measurement shows quite the same pattern, and the amplitude of the anomaly pattern is exaggerated compared with that of the surface ship measurement. This anomaly pattern suggests that the expected structural boundary exists at the central axis of Toyama Trough, not at the Deep Sea Channel just off the northern end of the Fossa Magna on land. If so, Toyama Deep Sea Channel should be a meandering channel formed by the faulting accompanied by the predominant movement of the structural boundary.

DISCUSSION

Negative free air anomaly at the Toyama Trough continues northwards along the topographic depression of the trough. This fact was pointed out in the past as an evidence of the new subduction along the eastern edge of the Japan Sea. However, at least in this survey area, no appearance of subduction has been found so far. In addition, the negative zone changes its direction in this part and continues to the inside of Toyama Bay (*according to* Tomoda and Fujimoto, 1982). The whole area of the Toyama Bay is considered to be suffering from the downgoing movement due to the active supply of sediments from Kurobe River or other rivers nearby, and so the subbottom structure is considered to be a deeply eroded basement topography superposed by thick sediment constructing a flat bottom topography. The result of gravimetry suggests that the sediment thickness at the centre of the Toyama Trough is about 1km larger than that at the both edges of the trough.

The same pattern of gravity anomaly and subbottom structure is observed in off-Urakawa and off-Miyazaki areas (*according to* Tomoda and Fujimoto, 1982; Segawa and Matsumoto, 1987). Both of them are located just on the landward side of a junction point of the trench. In the Bonin Trough, similar negative zone with larger amplitude exists, and the zone contacts with the Bonin Trench at the northern end of it, although no active subduction nor spreading is recognized. Judging from these facts, these negative gravity zone should be formed in the places where active sedimentation came to take place due to some reason and downgoing movement of the basement rocks continues in correspondence with the sedimentation.

ACKNOWLEDGEMENT : The authors would like to thank Prof. K. Kaminuma of National Institute of Polar Research Japan for his advice for gravimetry and mechanical problems of the gravimeter. Some other JAMSTEC researchres engaged in the sea bottom gravimetry during their research dives are also to be acknowledged. This work was supported by the Special Coordination Funds of the Science and Technology Agency of Japanese Government.

REFERENCES

Asada, A., Kato, S. and Kasuga, S. (1989). Tectonic landform and geological structure survey in the Toyama Trough, Report of Hydrographic Researches, No.25, 93-122.

Nakagawa, I. (ed.), (1983). Precise calibration of scale values of La Coste and Romberg gravimeters and international gravimetric connections along the Circum- Pacific zone (final report), pp117.

Nakamura, K. (1983). Possible nascent trench along the eastern Japan Sea as the convergent boundary between Eurasian and North American plates, Bull. Earthquake Res. Inst., Univ. of Tokyo, **58**, 711-722, (in Japanese).

Satomura, M., Otsuka, K., and Niitsuma, N. (1987). Precise gravity measurement on the floor of the Suruga Trough, and results of geological observations on the trough floor and its Shizuoka slope, Tecnical Reports of JAMSTEC, Special Issue, The 3rd symposium on deepsea research using the submersible "SHINKAI2000" system, 15-24, (in Japanese).

Segawa, J. and Matsumoto, T. (1987). Free air anomaly of the world ocean as derived from satellite altimetry data, Bull. Ocean Res. Inst., Univ. Tokyo, No.25, pp122.

Tomoda, Y., and Fujimoto, H. (1982). Maps of gravity anomalies and bottom topography in the Western Pacific, Bull. Ocean Res. Inst. Univ. of Tokyo, No.14, pp158.

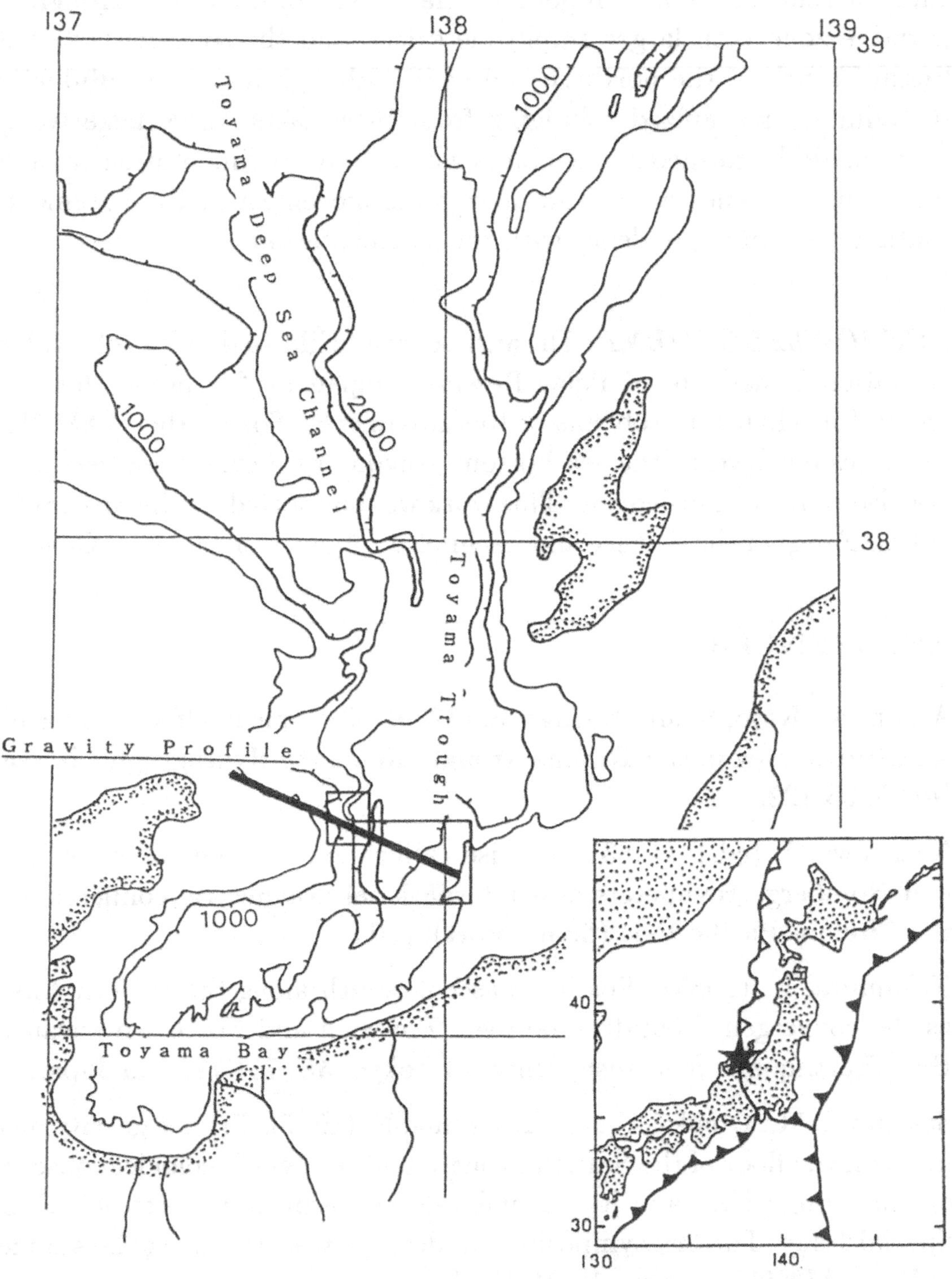

Fig.1 : Survey area of the sea bottom gravimetry in this work. Bold line in this figure is the gravimetric survey line.

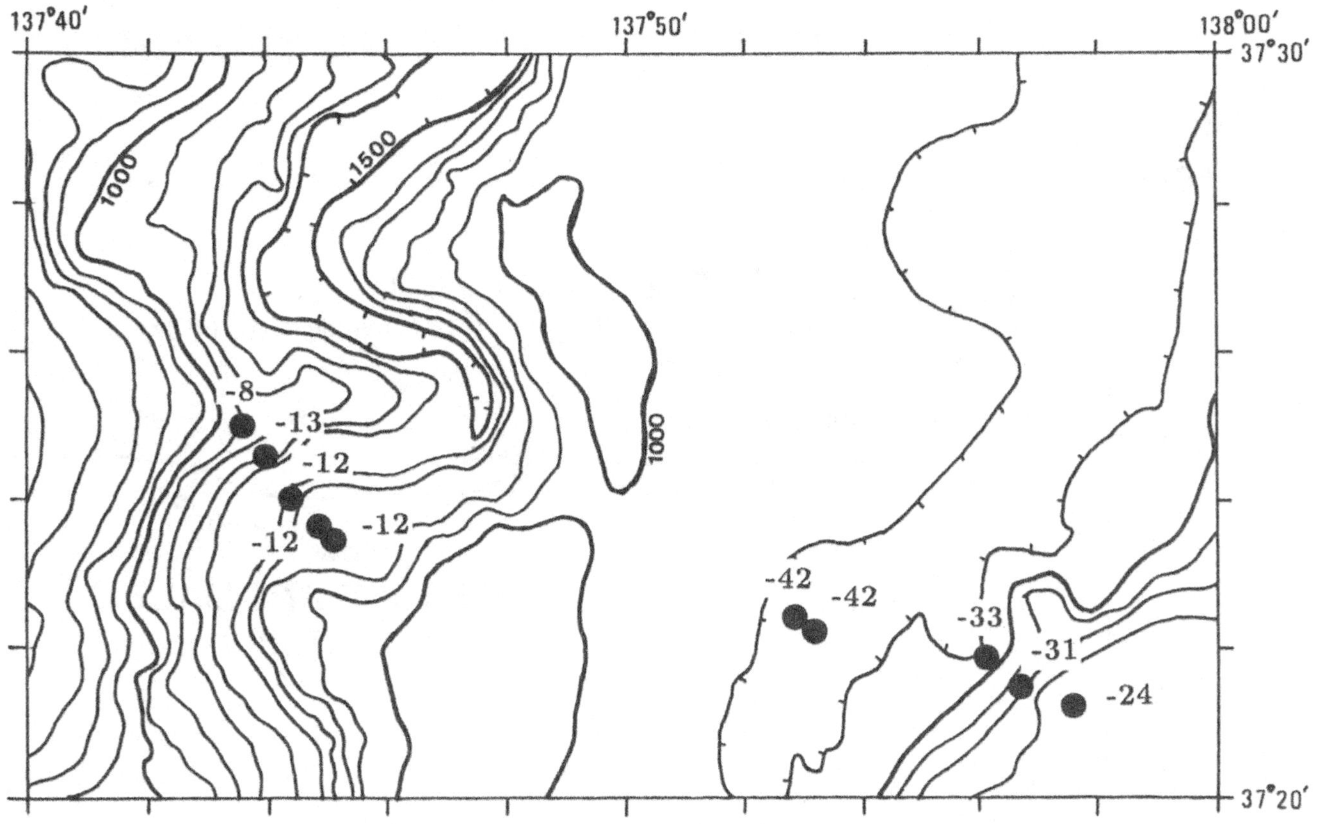

Fig.2 : Free air anomaly reduced to the level of the sea surface togethter with precise bathymetric map of Toyama Deep Sea Channel by SEABEAM system (this work; Asada et al., 1989). Numbers in bold letters are the value of free air anomaly on the Toyama Deep Sea Channel (measured in 1987) and on the Toyama Trough (measured in 1988). 2-dimensional terrain correction is conducted. Assumed density of the sea bottom material is 1.8 g/cc.

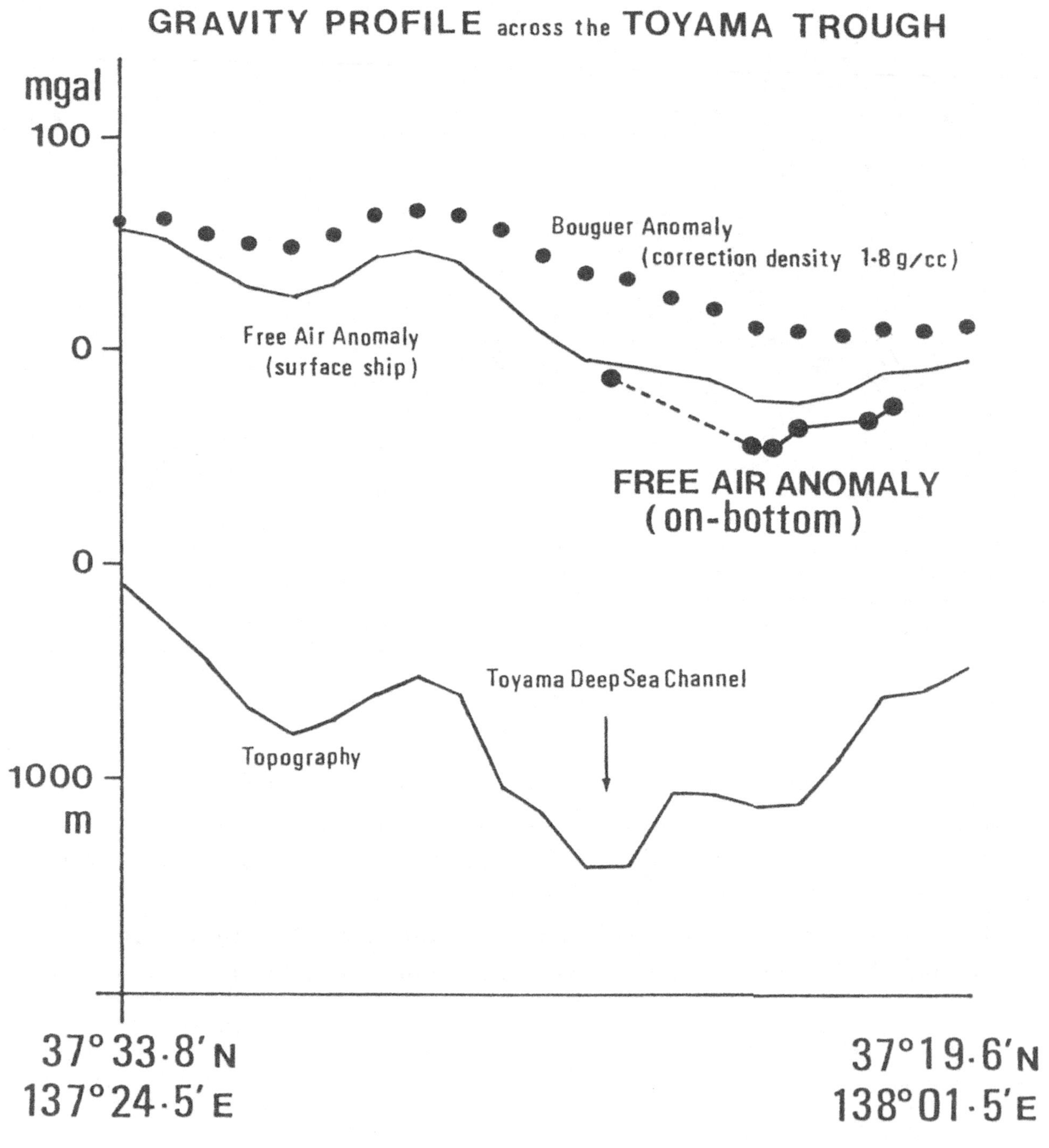

Fig.3 : Gravity and bathymetric profile across the Toyama Trough. Position of the survey line is shown in Fig. 1. This figure includes both the result of the sea bottom gravimetry (this work) and that of surface ship measurements by Hydrographic Department of Japan.

ADJUSTMENT OF MICROGRAVIMETRIC MEASUREMENTS FOR DETECTING LOCAL AND REGIONAL VERTICAL DISPLACEMENTS

Matthias Becker
Institut für Physikalische Geodäsie
Technische Hochschule Darmstadt, F.R.G.

ABSTRACT

Microgravimetric measurements in three different applications are ana-
lyzed with respect to different ways of their adjustment. The standard
least squares solution is compared to robust-M-estimation. The data of
a large scale geodynamic network in South America, of a local deformation net in Norway and of a very high precision indoor net are evaluated. The capabilities of robust estimation in these three cases with
rather different accuracies is investigated. It is found that, depending on the tuning constant, about 10 percent of the measurements
are affected by the robustification. The adjusted gravity values are
changed by one to two times their mean square error at maximum compared
to the ordinary least squares solution. Outliers are routinely detected
and their effect is reduced significantly. The complete procedure for
the adjustment of high precision gravimetry with observations in different epochs is presented.

INTRODUCTION

Microgravimetry can be a major tool for the detection of vertical displacements. In principle gravimetry gives ambigous results and has to
be aided by or used as aid for geometrical measurements like levelling
or satellite-positioning.

Presently we have two major projects where vertical displacements are
to be studied. The first is the subduction-zone at the western border
of South-America, where the recent uplift accompanying the subduction
is studied in an area of about 2000 km north-south and 800 km east-west
extension respectively. Long distances and various ways of transportation under severe observation conditions have to be used. Information
on geometrical changes of height is planned to be obtained from GPS-ob-
servations on part of the sites.

The second one is quite different. At the Blue-Lake artificial reservoir in Norway the enormous load exerted by 3000 mill. cubic meters
of water causes vertical displacement of the earth-surface. High precision gravimetric observations in combination with levelling and an additional GPS network are intended to monitor the changes. Here the extension of the profiles is only about ten to twenty kilometers and

well-controlled gravity observations are possible.

In a third application a small indoor network with gravity differences smaller than one mgal was observed with 13 different LaCoste G and D gravity meters, some equipped with electrostatic feedback. In these three cases different levels of accuracy can be expected due to the specific conditions.

However, a general problem in the adjustment of high precision gravity observations is the occurrence of outlying observations. In the sequel we will apply a robust-estimation method. This is done due to the fact that on one hand the gravimeters are very delicate instruments leading to a variety of error sources and on the other hand the observation scheme does not in every case allow a perfect modelling of the instrumental behaviour. Moreover the relatively few number of redundant observations and the possibility of gross errors can be expected to eventually distort the standard least squares estimation based on the assumption of normal distribution for the observations.

ROBUST ESTIMATION PROCEDURE

There are different strategies to handle the problems mentioned above. In geodesy "data-snooping" of Baarda (1968) and the τ-test of Pope (1975) are well known and established. These two statistical test procedures are based on assumptions for the distribution of the residuals and are used for a successive labelling and rejection of outlying observations.

Robust estimation methods were basically developed by Huber (1964) and Hampel (1973). The use in geodesy was initiated by Carosio (1979) and for the special case of gravimetric observations by Mäkinen (1981).

Borutta (1988) gives a more recent comprehensive survey of robust-methods with examples. The following derivation of formulas is based on the paper of Huber (1977) and Mäkinen (1981). We start with the standard Gauß-Markov Model:

$$y = Ax + \epsilon \tag{1}$$

with y = vector of n observations, A = design-matrix, x = vector of q unknown parameters and ϵ = vector of n true observational errors. We assume uncorrelated observations with equal weights, i.e.

$$Q_y = \sigma_0^2 I; \quad I = \text{unit-matrix} \tag{2}$$

In the standard least squares technique we are estimating the Best Linear Unbiased Estimates for the unknowns, which for normally distributed errors has the minimal variance.

$$\hat{x} = (A^T A)^{-1} A^T y \tag{3}$$

This is accomplished by minimizing the sum of squared residuals

$$\sum_{i=1}^{n} (y_i - \sum_{j=1}^{q} a_{ij}x_j)^2 = \min_x ! \tag{4}$$

In 1964 Huber introduced the M-estimators as a set of Maximum Like-
lyhood estimators where more generally an arbitrary function ρ of the
residuals is minimized.

$$M = \sum_{i=1}^{n} \rho(y_i - \sum_{j=1}^{q} a_{ij}x_j) = \min_x ! \tag{5}$$

or

$$M = \sum_{i=1}^{n} \rho(z_i) = \min_x ! \tag{6}$$

Depending on ρ we get different estimators, e.g. we get standard least
squares if $\rho(z) = z^2$, or the L_1-Norm for $\rho(z) = |z|$. In order to
minimize the effect of outlying observations we may choose

$$\rho(z) = \begin{cases} \tfrac{1}{2}z^2, & |z| \leq c \\[2mm] \tfrac{1}{2}c^2, & |z| > c \end{cases} \tag{7}$$

Taking derivatives we obtain the influence-function

$$\psi(z) = \frac{\partial \rho(z)}{\partial z} \tag{8}$$

$$\psi(z) = \begin{cases} z, & |z| \leq c \\[2mm] c \cdot \mathrm{sign}(z), & |z| > c \end{cases} \tag{9}$$

As can be seen from Fig. 1 and Fig. 2 the loss-function $\rho(z)$ is identical to that of least squares within the interval $- c \leq z \leq c$; outside
the interval it exhibits a linear increase in contrast to the quadratic
rise of the least squares influence function. The plot of the influence
function shows that compared to least squares the robust Huber estimator has a constant value outside the interval determined by the so called "tuning-constant" c. Practically this means that the residuals
greater than $\pm$ c have the same effect on the solution as those being $\pm$
c exactly.

There are other candidates of influence functions and associated
estimators, like the Hampel, Sine and Biweight-Estimator. Comparing and
testing robust estimators is often done using contaminated normal distributions. This means that part of the residuals are coming from a
standard normal distribution and part of them come from a "gross error
distribution" which can not be classified but is known to be symmetric.
The degree of contamination, usually taken to be between 1 to 10% is

determining the choice and tuning of the influence function. Besides the Hampel estimator, the estimators mentioned above are not well suited for geodetic applications because they are not identical to least squares for residual shorter than $\pm$ c. The Hampel estimator, where the influence of long residuals above a second tuning constant is attenuated to zero, is advantageous in case of a larger contamination. The extensive comparison of robust estimators made by Andrews et al (1972), indicates, that the Huber-estimator is more likely to be adequate for our application with relatively small sample sizes.

For the implementation of the robust estimation we chose the W-algorithm of Dutter (1975) where scale and parameters are estimated by iteratively applying re-weighted standard least squares algorithms with modified weights for the observations.

1. The procedure starts with a zero order approximation e.g. with a least squares estimate for $\hat{\sigma}^o$, $\hat{x}^o$.

2. First estimates for the residuals are obtained as

$$\hat{e}_i^m = A\hat{x}^m - y \tag{10}$$

3. With these the new estimate for the scale is

$$(\hat{\sigma}^{m+1})^2 = \frac{\beta}{n-q} (\hat{\sigma}^m)^2 \cdot \sum_{i=1}^{n} X_i \left(\frac{e_i^{\,m}}{\hat{r}^m} \right) , \tag{11}$$

with

$$X(z) = 2z\psi(z) - \rho(z) = \begin{cases} z^2 & |z| \leq c \\[2mm] c^2 & |z| > c \end{cases} . \tag{12}$$

By using $\hat{\sigma}^m$ in (11) the point c is kept fixed throuout the iterations in spite of changing $\hat{\sigma}$. The factor β is used to get an asymptotically unbiased estimator for the scale and is defined as:

$$1/\beta = E \{ x(u) \} \tag{13}$$

with u being normally distributed.

4. Now the elements of the diagonal weight-matrix w can be computed as

$$w_{ij}^{m+1} = \begin{cases} \dfrac{\psi(e_i^m/\hat{\sigma}^{m+1})}{(e_i^m/\hat{\sigma}^{m+1})} & i=j \\[4mm] 0 & i \neq j \end{cases} \tag{14}$$

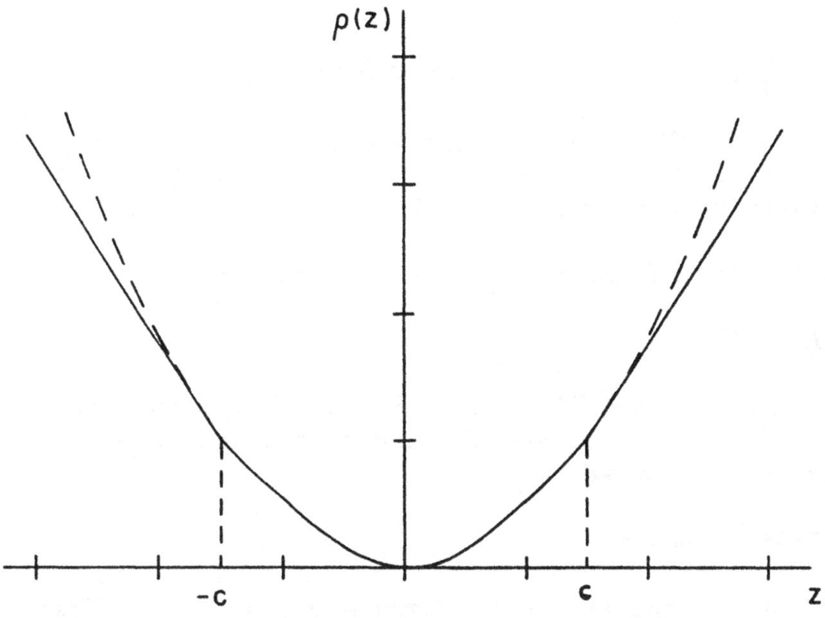

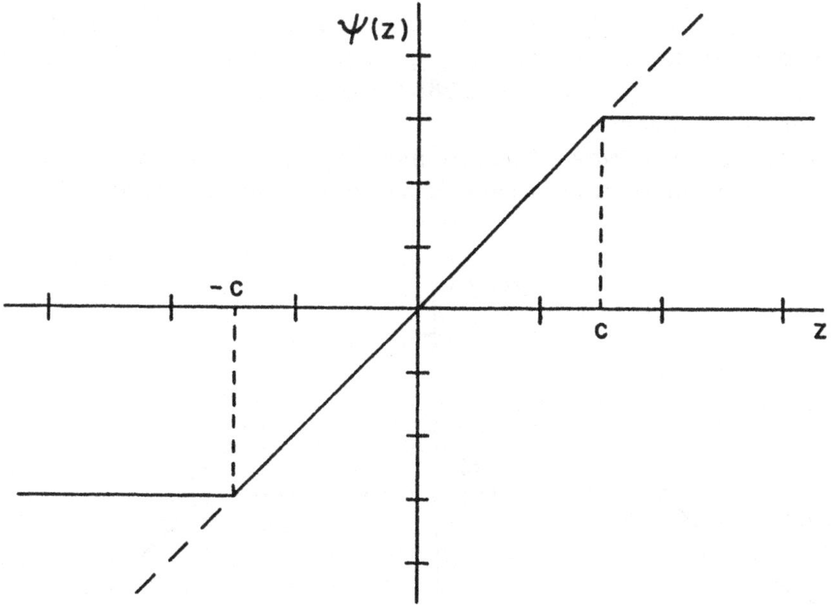

Fig. 1 Influence and ψ-function of the Huber type estimator (solid-
-lines) and of the standard Least Squares estimator (dashed
lines)

so that

$$
w^{m+1}_{ii} = \begin{cases} 1 & e^m_i \leq \hat{\sigma}^{m+1} \; c \\[2mm] c \; (\hat{\sigma}^{m+1}/e^m_i) & e^m_i > \hat{\sigma}^{m+1} \; c \end{cases} \tag{15}
$$

The function is plotted in Fig. 2.

5. A new estimation of the parameters is given by

$$
\hat{x}^{m+1} = (A^T W A)^{-1} A^T W y \tag{16}
$$

If the change between $\hat{x}^{m+1}_j$ and $\hat{x}_j^m$ is small enough the iteration can be stopped and the covariance matrix of the parameters can be estimated by

$$
Q_x = (\hat{\sigma}^{m+1})^2 \; (A^T W A)^{-1} \tag{17}
$$

There are other choices how to compute Q_x, see e.g. (Hill, 1979). If the changes in $\hat{x}$ are too big further iterations can be computed starting at step 2 again.

We used the W-algorithm with our standard least squares adjustment program due to convenience. If computing time is a critical factor there are other choices, e.g. Borutta (1988) gives different strategies and comparisons of performance.
  The tuning constant c was choosen as 1.5 which is adequate for slightly contaminated observations with a degree of contamination of 1–10%.

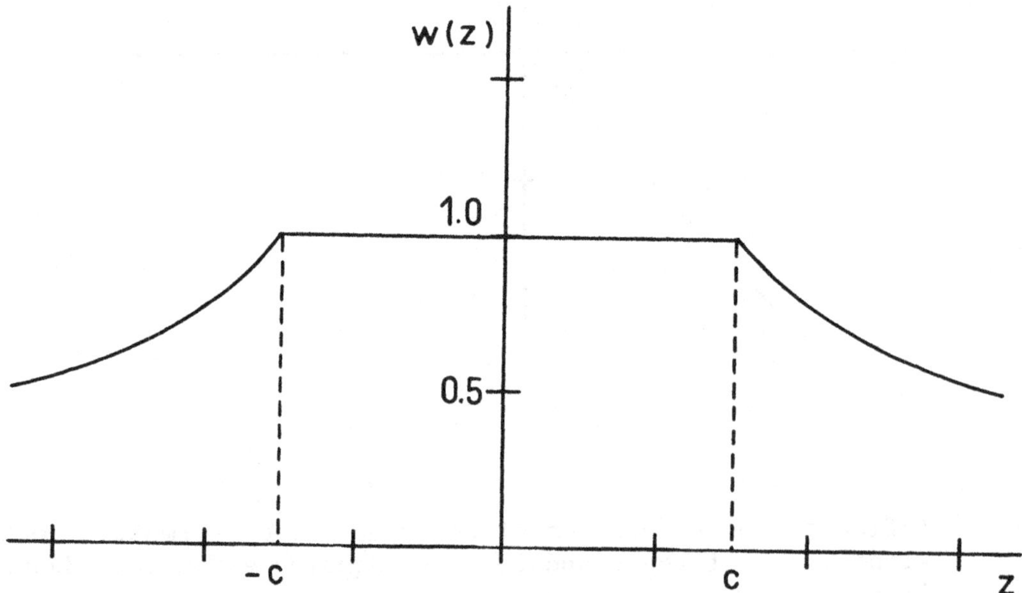

Fig. 2. Weight function of the Huber estimator

NUMERICAL RESULTS

The robust estimation was tested for the adjustment of repeated campaigns in South-America, the Blue-Lake project and the microgravimetric
net at the 1985 intercomparison of Absolute Gravimeters in Paris. The
functional model uses the gravimeter readings as observables and ad-
justs gravity values and additional parameters for drift, scale-func-
tion, periodical errors and off-sets. A detailed description can be
found in (Becker, 1984). All parameters are tested for their significance after the adjustment to find the optimal number of parameters,
see (Becker, 1981) for details.

The final values of the iteratively improved gravity values of different epochs are then compared and the statistical methods described
e.g. in (Niemeier, 1985) are applied to allow a judgement about the
significance of the variations, see also (Aksoy et al., 1988) for details of the analysis employed here. The adjustment-procedure is explained in the flow chart in the appendix.

Table 1. Convergence of Robust adjustment for the Blue-Lake Campaign
1988, iteration limit 0.3 μgal change in adjusted gravity
values. n = number of observations, q = number of parameters, nl = number of long residuals at first and last
iteration step, all values in μgal.

n	q	nl	Mean Square Error of Unit Weight (m.s.e.) for iterations				
			0	1	2	3	4
all instruments							
296	93	28/27	17.63	12.93	12.28	12.10	12.01
all instruments without helicopter measurements							
225	70	21/21	11.89	10.33	10.03	9.95	–
G258							
93	45	7/7	21.09	14.07	12.52	12.18	–
G563							
101	48	6/6	11.75	10.69	10.28	10.16	–
G688							
102	48	8/8	8.78	8.13	8.04	8.01	–

Table 1 gives a survey of robust estimation results for the Blue Lake campaign 1988 with 3 LCR gravimeters. Gravity measurements were conducted using helicopter transport on two days. Due to bad weather conditions and problems with insulation the performance was rather poor on these days. Large drifts and occasionally jumps deteriorated the observations, furthermore due to limited flying time some sites could only be observed once. The difference in accuracy can be seen from Table 1 if one compares the results of the initial adjustments of all instruments with and without helicopter. The error of the latter is only about 70% of the first one. However, for the investigation of the robust estimation procedure we used all available observations, including the helicopter days. In the common adjustment the mean square error of unit weight drops from the initial value of the least squares solution of 18 μgal to about 12 μgal for the robust estimator after convergence.

Fig. 3. shows the changes in gravity values. The robust estimator succeeds in labeling the gross errors of one instrument (with residuals >100 μgal) on the helicopter days. Weights of 27 readings are changed. As we have no "true" value for comparison we judged the improvements of the robustified estimation by comparing the inter-instrumental differences.
Especially G258 is improved considerably by the iterated estimation. The other instruments show only minor improvements in spite of the fact that the number of long residuals is about 6 to 9% in all cases. The r.m.s discrepancy between instruments is reduced by about 1% for G258 versus G688 and about 15% for G563 versus G688.
This is only a small improvement, if any. The reason lies in the poor design of the observations. Major discrepancies occur on the stations with no redundant observations of single instruments and these happen to be those with gross errors. The use of several instruments simultaneously and the combined adjustment resolves the bad readings. About 50% of the large residuals in single and combined adjustment are different.
In case of our large geodynamic network in South America (Becker et al., 1989) the situation is quite different. There are much bigger gravity differences of up to 2 gal and long distances with unfavourable observation conditions. Especially the modelling of drift and systematic instrumental errors is difficult and not perfect. In Table 2 the results of the robust-estimation are compared to those of the least squares solution. Here we used a combined adjustment of both epochs. The maximum change in some adjusted gravity differences is larger than their error obtained in the adjustment. There are some real large residuals which may indicate gross errors, but there is no real blunder as in case of the Blue-Lake results.

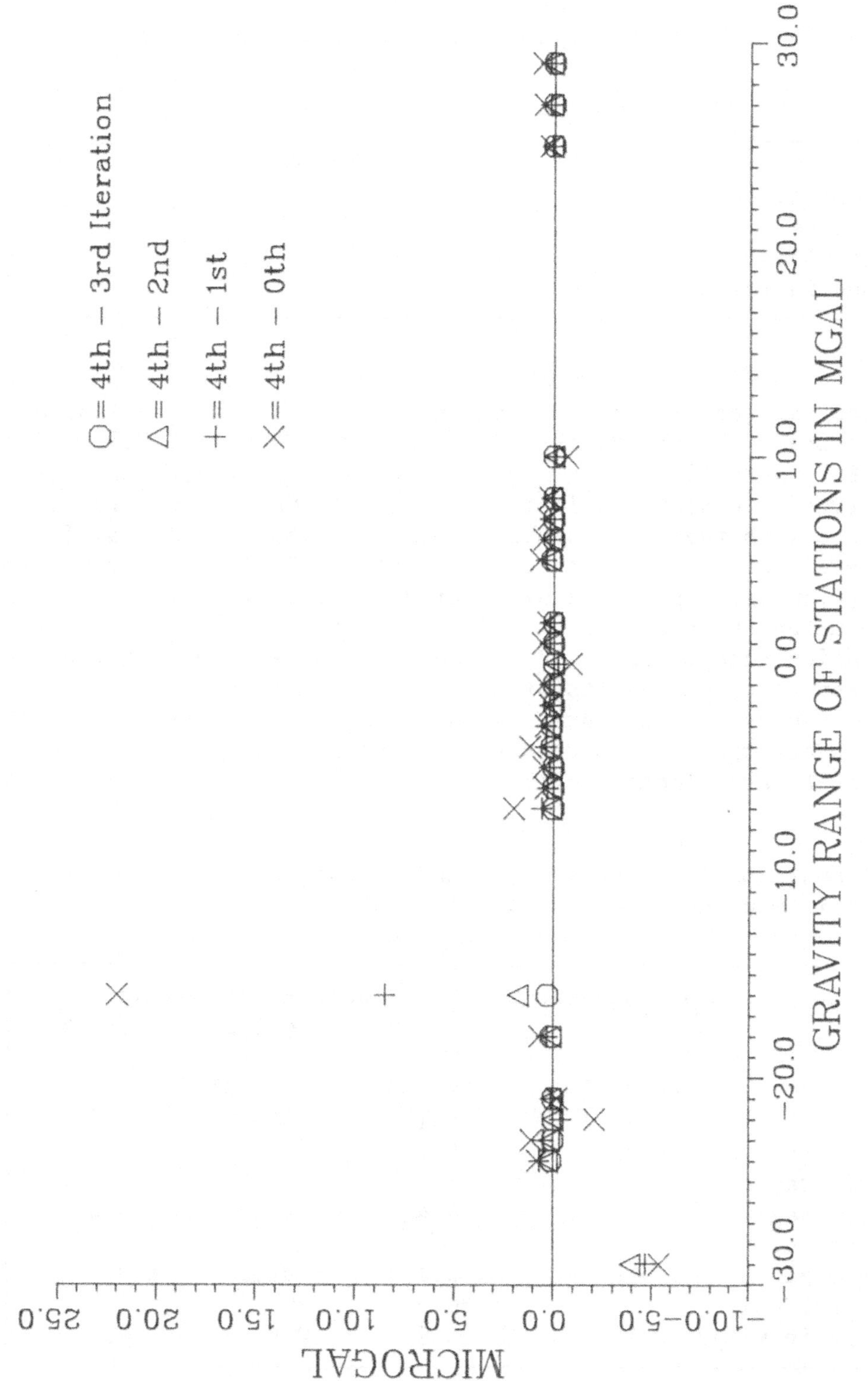

Fig. 3. Gravity differences between iterations

Table 2. Comparison of least squares and robust estimation for
South-America net, all values in μgal.

	Least Squares	Robust-M- Estimator
n/q	1743/504	1743/504
nl	–	182
m.s.e.	24.19	20.97
m.s.e. of gravity values (average)	13.61	12.20
max.difference to least squares	–	± 10
maximal residual	90.	140.
max change in weight	-	.8

Almost the same conclusion but at a totally different level of accuracy can be found for the Paris 1985 micronet (Boulanger, Faller and Groten, 1986). Here 13 gravity meters, partly with electrostatic feedback, observed at 6 stations inside a building. Under these favourable condition and with the large number of observations gravity differences could be determined as accurate as 0.6 μgal. As Table 3 shows the robust estimation again gives results which are different from the least squares solution but are still within the errors of the adjustment. The robustification changes about 12 percent of the weights. For the sake of comparison also the results of different solutions with various values of the tuning constant are listed in Table 3. Due to the large number of observations the percentage of modified observations changes but the results are almost equal.

Table 3. Comparison of least squares and robust estimation for the
Paris micronet, all values in μgal.

	Least Squares	Robust-M- Estimator $c=2.0$	$c=1.8$	$c=1.5$
n/q	640/131	640/131	640/131	640/131
nl	–	31	43	86
m.s.e.	4.52	4.25	4.13	3.84
m.s.e. of gravity values (average)	0.43	0.42	0.42	0.42
max.difference to least squares	–	± 0.1	± 0.15	± 0.2
maximal residual	20.	20.	20.	20.
max change in weight	-	0.7	0.6	0.8

CONCLUSIONS

The HIPGRAD package containing correction and robustified adjustment
with statistical tests for significance of high precision gravimeter
measurements as well as statistical tests for determining gravity changes between epochs has been used with different data sets. The robust M
estimation improved the results by detecting the gross errors and other
outlying observations. Sensitivity of the least squares method against
outliers is reduced. By analyzing the weights and the residuals in combination with tests for the significance of parameters the functional
model can be improved.

In spite of the rather automated procedure care has to be taken in
order to detect deficiencies in the design of measurements which can
not be checked by the robust estimation. It is still recommended to use
as much different gravimeters in parallel as possible in order to prevent biased results due to unmodelled systematic errors or effects
which can not be introduced to the model.

The choice of the tuning constant should be made according to the degree of contamination, but seems not to be critical. It should be kept
as large as possible in order to reduce the special treatment to as few
observations as possible. A benefit from the robust adjustment is the
inclusion of all observations so that anybody can see the changes made.
If furthermore the original as well as the iterated mean square errors
are reported, additional information on the quality of the measurements
can be deduced.

Further theoretical and practical experience is needed. The optimal
computation of the covariance matrix of the unknowns could be investigated, as well the choice of correction factors and tuning constants.
The combination of different instruments with their possibly different
inherent accuracies leads to the question of an optimal combination of
robust estimation with variance-component estimation.

ACKNOWLEDGMENT. I am thankful to J. Mäkinen who brought the topic of
robust-estimation to my attention and with the statement "It looks aw-
ful but really isn't" in his original paper encouraged me in the application.

REFERENCES

Aksoy, A., M. Becker, H. Demirel, E. Groten, W. Hönig (1988)
        Präzisionsschweremessungen zur Überwachung von Vertikalbewe-
        gungen in der nordanatolischen Verwerfung. — Deutsche Geodä-
        tische Kommission, Reihe B, 287, 1988, München
Andrews, D.F., P.J. Bickel, F.R. Hampel, P.J. Huber, W.H. Rogers, J.W.
        Tukey (1972) Robust Estimates of Location, Survey and Advances.
        Princeton University Press, Princeton, New Jersey
Baarda, W. (1968) A testing procedure for use in geodetic networks,
        Netherlands Geodetic Commission, Publications on geodesy, 2:5

Becker, M. (1984) Analyse von hochpräzisen Schweremessungen. – Deutsche
        Geodätische Kommission, Reihe C, 294, 1984, München
Becker, M., S. Bakkelid, E. Groten, B.G. Harsson, A. Midsundstadt
        (1988) Investigation of crustal deformation induced by loading
        of an artificial lake - The Blasjö-Project. – Bull. D.'Inf. BGI
        No. 62, 1988, Toulouse
Becker, M., M. Araneda, E. Groten, O. Hirsch, T. Knöll and S. Ostrau
        (1989) First results of repeated gravity measurements in the
        southern central andes – Zbl. Geol. Paläont. Teil I, 1989 (5/6),
        Stuttgart
Borutta, H. (1988) Robuste Schätzverfahren für geodätische Anwendungen.
        Schriftenreihe des Studiengangs Vermessungswesen, Universität
        der Bundeswehr München, Heft 33
Boulanger, Y.D., J. Faller and E. Groten (Eds) (1986) Results of the
        second international camparison of absolute gravimeters in
        Sevres 1985, Bull. D.'Inf. BGI No. 59, 1986, Toulouse
Carosio, A. (1979) Robuste Ausgleichung. Vermessung, Photogrammetrie
        und Kulturtechnik, 77, 293-297
Dutter, R. (1975) Numerical solution of robust regression problems:
        computational aspects, a comparison. Fachgruppe für Statistik,
        ETH Zürich, Research Report No 7, Zürich
Förstner, W. (1978) Die Suche nach groben Fehlern in photogrammetri-
        schen Lageblöcken, Deutsche Geodätische Kommission, Reihe C,
        240, München
Hampel, F.R. (1973) Robust estimation: a condensed partial survey, Z.
        Wahrscheinlichkeitstheorie verw. Geb. 27, 87-104
Hill, R.W. (1979) On estimating the covariance matrix of robust re-
        gression M-estimates. Commun. Statist. - Theor. Meth. A8 (12),
        1183-1196
Huber, P.J. (1964) Robust estimation of a location parameter, Ann.
        Math. Statist., 35, 73-101
Huber, P.J. (1972) Robust statistics: a review, Ann. Math. Statist.,
        43, 1041-1067
Huber, P.J. (1977) Robust methods of estimation of regression coef-
        ficients, Math. Operationsforsch. Statist., Ser. Statistics, 8,
        41-53
Mäkinen, J. (1981) The treatment of outlying observations in the ad-
        justment of the measurements on the Nordic land uplift gravity
        lines, unpublished manuscript, Helsinki
Niemeier, W. (1985) Deformationsanalyse. – "Geodätische Netze in
        Landes- und Ingenieurvermessung II", H. Pelzer (Ed.) 1985,
        Stuttgart
Pope, A.J. (1975) The statistics of residuals and the detection of
        outliers, XVI. General Assembly of the IAG, Grenoble

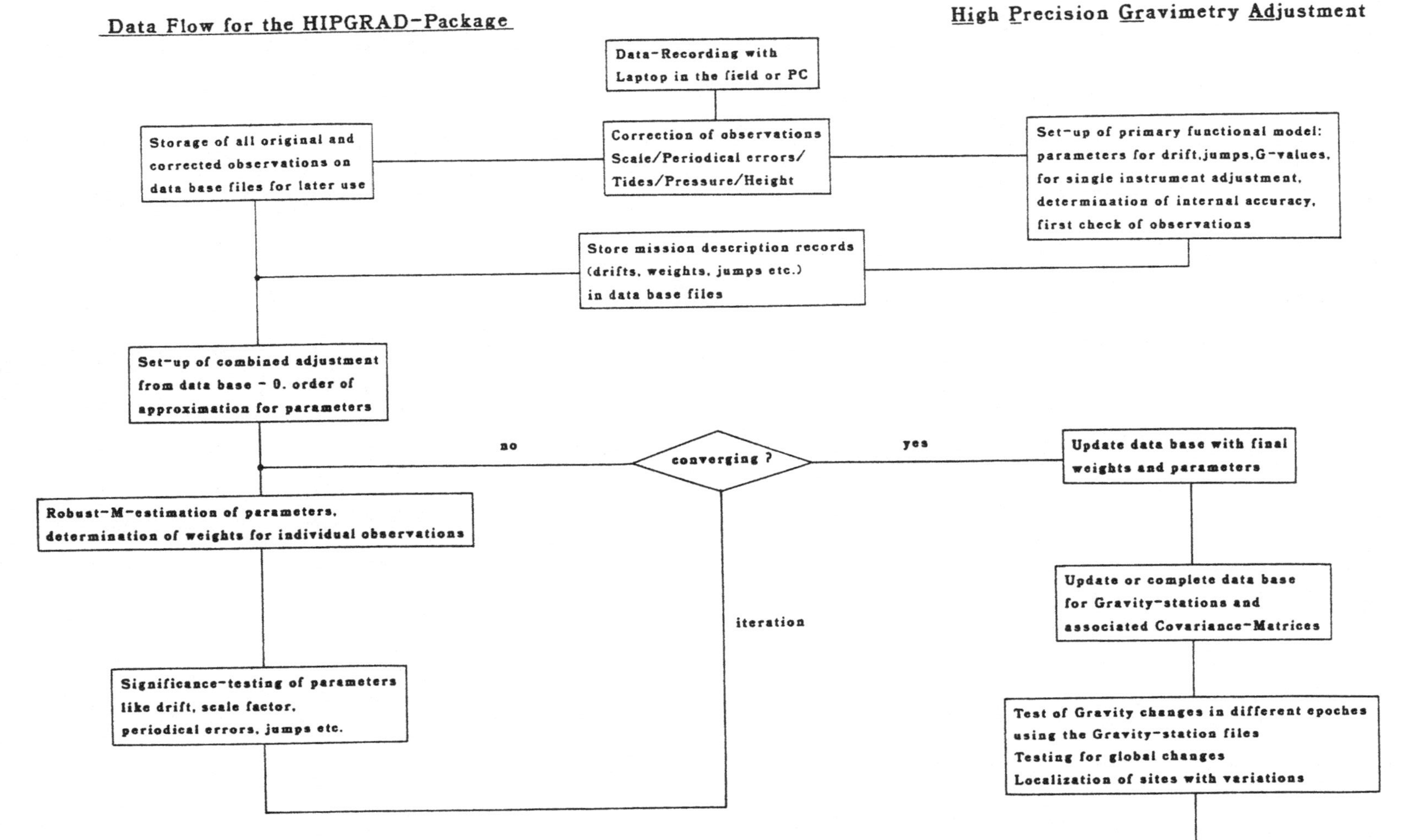

Appendix A. Flow chart of data evaluation

SECULAR GRAVITY VARIATIONS ALONG THE NORTH ANATOLIAN FAULT

H.Demirel
Yildiz University Istanbul, Turkey
C.Gerstenecker
Technische Hochschule Darmstadt, FRG

Abstract: Within the cooperative Turkish-German project on earthquake prediction research a gravimetric network along the western part of the North Anatolian fault zone (NAFZ) between Adapazari and Bolu was set up. Applied observation and adjustment procedures are described and results of the first two observation campaigns (1988 and 1989) are given. Significant gravity changes up to 450 [nm sec^{-2}] between northwestern and southeastern stations are obtained. Vertical crustal movements or other physical processes however must be excluded as reasons. On the contrary we assume that the gravity changes reflect not well understood instrumental problems.

INTRODUCTION

Perhaps the most important idea behind the interdisciplinary Turkish German project on earthquake prediction research (Ergünay et al.,1989) is to combine all known methods of earthquake precursors research within a well defined project in the same region. As area of investigation the Mudurnu valley at the North Anatolian Fault Zone (NAFZ) in Turkey between Adapazari and Bolu was choosen.
Control of the gravity field in this area due to secular gravity changes is part of the project. For that purpose a regional gravimetric network was set up. The network (Fig. 1) consists of 22 stations and connects two further already existing networks (Hipkin, 1987 and Becker et al., 1985). At each station 2 observation sites are marked by special bench marks. The distances between both sites are between 20 and 400 m. Height differences between the sites are determined by geometric levelling.
Basis station of the network is the station "Yigilca", where repeated coordinate determinations with mobile laser ranging systems in the frame of the WEGENER-MEDLAS project were carried out.
More details about the network are given in (Demirel et al.,1989)

OBSERVATION TECHNIQUES

The observations were carried out with relative spring gravimeter LaCoste Romberg, model G. The gravimeters used in the different field campaigns are shown in Table 1.
The gravity differences are measured forward and backward along lines. In minimum each gravimeter was observed four times at each observation site.
In addition to the gravimeter readings air temperature, air pressure, battery voltages and the height of the gravimeters above bench marks are

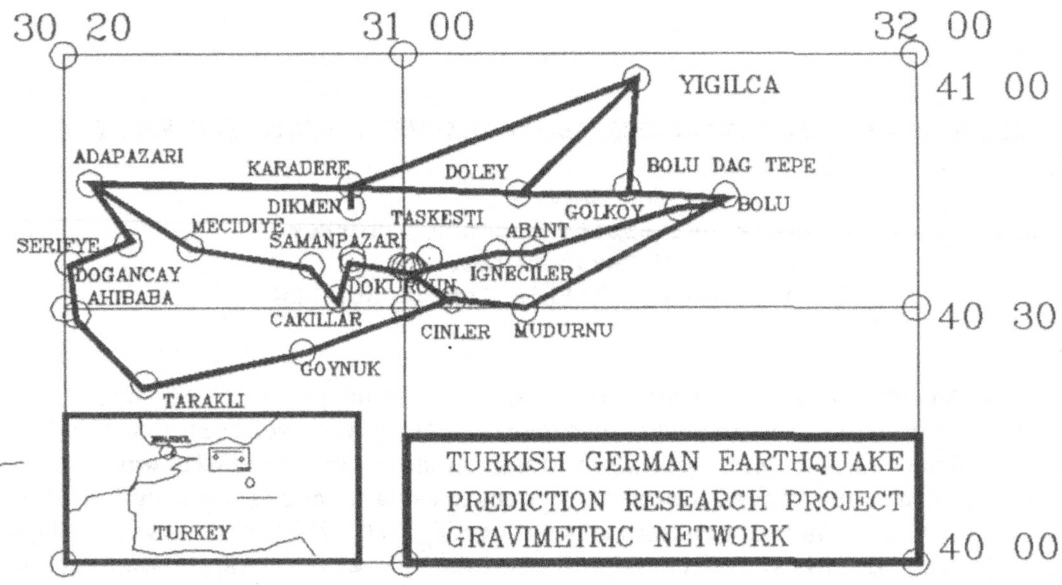

Fig. 1 Gravimetric Network

observed. All observations were taken with an automatic data logging system and stored in the data bank system "FELDGRAV" as described by
(Czuczor et al., 1989). Each measurement was reduced and controlled in
the field immediately. In case of great drift rates the observations were repeated.
The gravimeters were transported by cars in their original alu-cases as
delivered by the manufacturer. Bad roads and great temperature variations are reflected in great irregular drift rates of the gravimeters as
shown in fig. 2.
For the observation of the whole network between 12 and 16 days are
necessary. Till now two observation campaigns were carried out :
- September 1988
- July 1989

During the first campaign the network was set up. All observations were
carried out with four gravimeters by only one observer.
For the second campaign in July 1989 only two gravimeters were
available. Both instruments were read by two observers at each observation site. During this campaign the LaCoste-Romberg gravimeter LCR-G
258 was established with a Harrison-Sato feedback. For comparison this
gravimeter was observed as a third instrument "LCR-G 258$^+$" using the
feedback system.

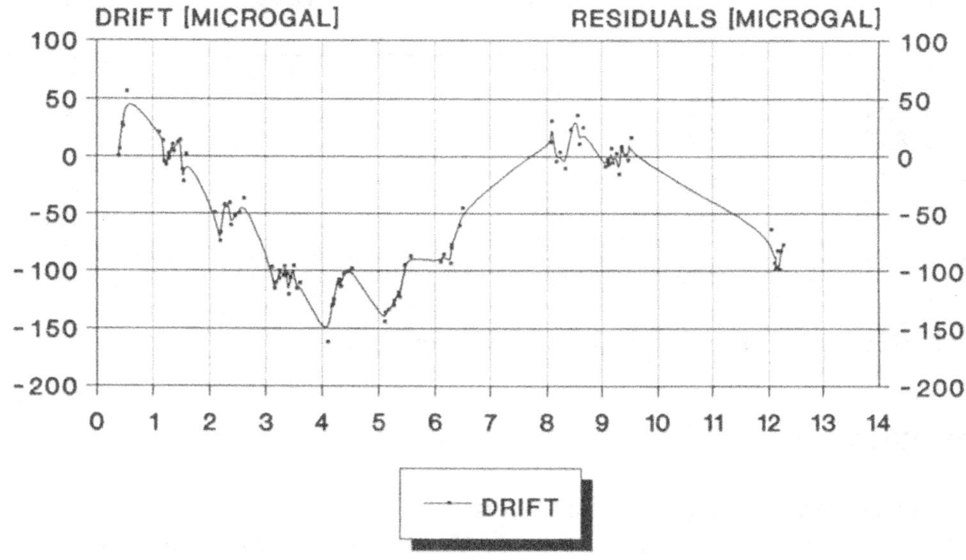

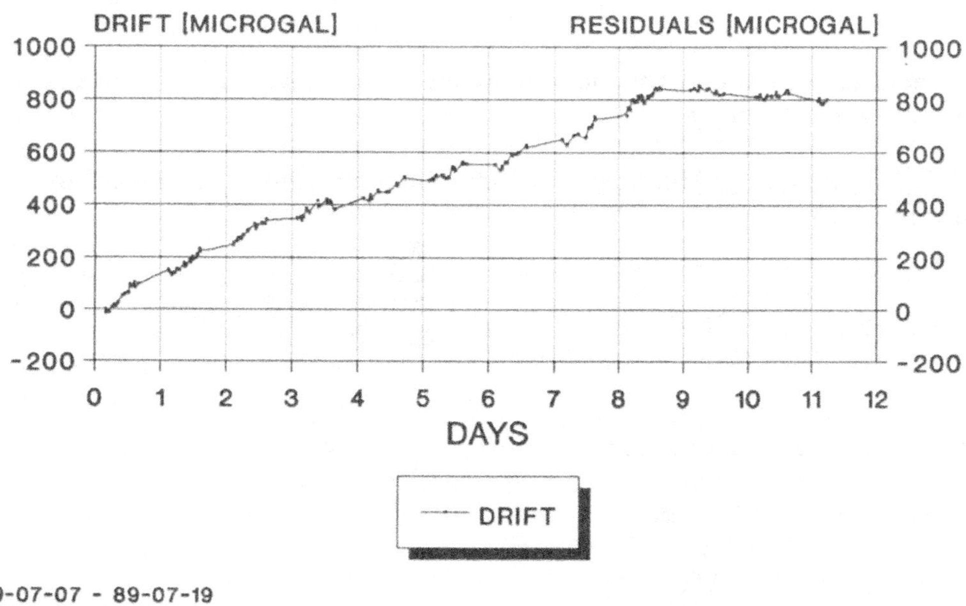

Fig. 2 Example of Gravimeter Drift

Table 1: Gravimeters and their mean square errors of unit weight m_0

Gravimeter	Number of observations		Mean square error of unit weight [nm sec^{-2}]	
Epoch	1988	1989	1988	1989
LCR-G 258	98	271	10	8
LCR-G 563	91	262	12	10
LCR-G 839	82		13	
LCR-G 874	83		14	
LCR-G 258[+]		260		7
Total	354	797	16	12

[+] LCR-G258 with Harrison-Sato Feedback

DATA ANALYSIS

Following <u>reductions</u> were applied:
- calibration factors from manufacturer's calibration table
- linear scale factor of the International Gravity Standardization Network 1971 (IGSN 71)
- periodic spindulum errors (if available)
- gravity changes due to different instrumental height above the bench mark
- earth tidal gravity using Büllesfeld's(1985) tidal development and the elastic Wahr earth model as modified by (Dehant et al, 1986)
- gravity changes due to air pressure changes.

No reductions are considered concerning rainfall, soil moisture and ground water changes.

The <u>least square adjustment</u> of the observations is based on observation equation (1).

$$v_{i,j,k} = G_j + S_k * Z_{i,k} + N_{1,k} + \sum^{m} D_{m,k} * t_i^m - A_{i,j,k} \qquad (1)$$

with

    i = Index of observations
    j =   "   " station
    k =   "   " gravimeter
    1 =   "   " niveau constant
    p =   "   " drift polynom
    m = degree of drift polynom

    G = unknown gravity value
    S = unknown scale factor
    N = unknown constant drift term (jump)

D = unknown polynomial drift coefficient

t = observation time
Z = gravimeter reading in scale units
A = reduced gravimeter reading in nm sec^{-2}

For diagnosis purposes the observations of each instrument during each field campaign were adjusted separately.The lowest mean square errors of the unit weight m_0 was obtained, if daily linear drift polynomials and daily constant drift terms are introduced in the adjustment model (Table 1).

Observations with residuals exceeding the $3*m_0$ threshold of the mean square error of the unit weight (as given in Table 1) were cancelled. Finally for each observation epoch a free network adjustment with all gravimeters together was carried out whereby the scale factor of the gravimeter LCR-G 258 was kept fixed. Thereby all observations were introduced with the same weight $P = 1$.

As final results the gravity values $G_{j,1,E}$ at the observation site 1 of the station j for the epoch E and their standard deviations m_G were obtained.

RESULTS AND DISCUSSION

The observations carried out 1989 show better inner accuracy. Reasons for that is the greater number of observations and the less number of gravimeters. That coincides with the fact, that crowing number of used gravimeters does not improve the inner accuracy of a network due to unknown systematic errors of each individual gravimeter (see f.e. LaCoste et al,1989). We believe therefor, that the outer accuracy of both observation campaigns is of the same order.

Significant scale factor changes with time were not detected.

Gravity changes $DG_{j,1}$ at the observation site 1 (1 = 1,2) of the station j between the epochs 1988 and 1989 are computed from (2)

$$DG_{j,1} = G_{j,1,1989} - G_{j,1,1988} \qquad (2).$$

The mean gravity changes DG_j

$$DG_j = 1/2 *(DG_{j,1} + DG_{j,2}) \qquad (3)$$

are shown in Fig. 3. Their standard deviations $m_{DG(j)}$ are estimated according

$$m_{DG(j)} = 1/2 * \sqrt{\sum^{j,1,E} m^2_{G(j,1,E)}} \qquad (4).$$

If a confidence level of $2 * m_{DG(j)}$ (γ = 95.5 %) is assumed statistical significant gravity changes in the northwestern and southeastern part of the network are found. Insignificant changes are estimated for all stations in the middle of the network.

Interpretation of these results as vertical crustal movement is

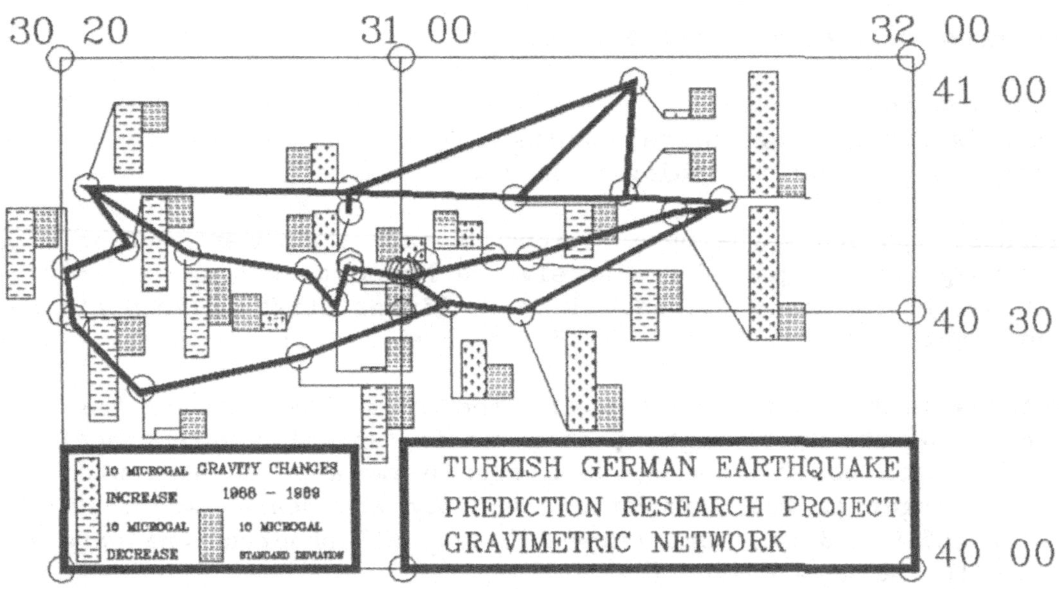

Fig. 3 Gravity changes between September 1988 and July 1989

unlikely. The gravity changes of about 450 nm sec^{-2} postulate crustal uplift between the station Adapazari and Bolu of more as 40 cm within 11 months, if a Bouguer gradient of 10 nm sec^{-2}/mm is assumed.
Similarly these gravity changes cannot be earthquake precursors. Gravity changes caused by dilatancy effects before an earthquake are very small and can hardly be observed with normal relative spring gravimeters as shown by (Lambert et al, 1989). Moreover such trends must be confirmed through more observations, which still now are not available.

Ground water changes, soil moisture and rainfall can cause gravity changes of about 100 - 200 nm sec^{-2} . However from ground water observations at the stations Samanpazari, Dokurcun, Taskesti, Igneciler and Abant (Kümpel et al., 1989) we know, that the ground water table is changing in the middle part of the network within one year not more as 10 cm in a depth > 20 m. Therefor the probability for gravity changes of observed order due to ground and surface water effects is very low.
The most likely reasons are seen in instrumental and adjustment problems. Gravimeter drift as shown in Fig. 2 is only in the long term part smooth and piecemeal linear. The smooth regular long term drift is superposed by irregular short term distortions (up to $\pm$ 400 nm sec^{-2}), which hardly can be fitted by linear drift coefficients. Short period drift is not only connected with the creep of the gravimeter spring; it can be caused by levelling problems, temperature changes, hysteresis or other unknown instrumental problems. From repetition measurements in the field we know that outliers as shown in Fig. 2 are real,repeatable observations, which are not caused by reading errors. Test computations with varying drift polynomials and jump eliminations models have produced gravity changes in the order of 200 nm sec^{-2} and more. Deviations

from the assumed drift model and neclected jumps are generating significant apparent gravity changes without improving the model fit.
Concluding at the moment no final answer is possible, if the observed
gravity changes are real physical changes or generated by insufficient
modelled instrumental problems. We are affraid that the outer accuracy
of the network is much lower as represented in Table 1 by the mean square error m_0 of the unit weight. The results of further field campaigns
must help to solve this problem.

Acknowledgements

The authors thank Nevcat Büyükköse, Cetiner Türkum, Demir Altin and Jo-
hannes Krüger who have participated partially at the different field
campaigns. Gravimeters were loaned by the Institute for applied Geodesy,
Frankfurt, the Institute of Geophysics, ETH Zürich and the Institute of
Geophysics, Christian-Albrechts Universität Kiel. Financial support was
given by the "Deutsche Forschungsgemeinschaft".

REFERENCES

Becker,M., Aksoy,A., Demirel,H. and E.Groten (1985): High Precision
    Gravity Measurements Across the North Anatolian Fault Zone, Bulletin
    D'Information, No. 57, Bureau Gravimetrique International, Toulouse
Büllesfeld, F.J. (1985) : Ein Beitrag zur harmonischen Darstellung
    des gezeitenerzeugenden Potentials, Deutsche Geodätische Kommission,
    Reihe C, No. 314, München
Czuczor,E., Falk,H.,C.Gerstenecker (1989): Automation in der
    Feldgravimetrie, Zeitschrift für Vermessungswesen, Vol. 114, pp 259-
    268
Dehant,V., B.Ducarme (1986): Comparison between the theoretical and
    observed tidal gravimetric factors, Bulletin D'Information, No.59,
    Bureau Gravimetrique International, Toulouse
Demirel, H, C.Gerstenecker (1989): Gravimetric Levelling Along The
    North Anatolian Fault Zone Between Adapazari and Bolu, Earthquake
    Prediction, J.Zschau and O.Ergünay (eds.), Christian-Albrechts
    Universität, Kiel
Ergünay,O., J.Zschau (1989) : The Turkish-German Project on
    Earthquake Research: An Overview, Earthquake Prediction, J.Zschau
    and O.Ergünay (eds.),Christian-Albrechts Universität, Kiel
Hipkin,R (1987): Private Communcation
Kümpel,J., N.Büyükköse (1989): Well Level Data at Five Multiparameter
    Stations in the Mudurnu-Abant Valley, Turkey, Earthquake Prediction,
    J.Zschau and O.Ergünay (eds.),Christian-Albrechts Universität, Kiel
LaCoste,L.B., H.P.Valliant (1989) : Gravity Meter Calibration At
    LaCoste And Romberg, paper presented at General Meeting of IAG,
    Edinburgh
Lambert,A., D.R.Bower (1989) : Constraints On The Usefulness Of
    Gravimetry For Detecting Precursory Crustal Deformations, paper pre-
    sented at 25th General Assembly of IASPEI, Istanbul

REGULARISING TUNNELLING ALGORITHM IN NON-LINEAR GRAVITY PROBLEMS - A NUMERICAL STUDY

R. G. S. Sastry[*] and P. S. Moharir[**]

[*] Department of Earth Sciences, University of Roorkee, Roorkee- 247667 (India)

[**] Presently, Visiting Scientist, National Geophysical Research Institute, Hyderabad-500007 (India)

ABSTRACT

Non-linear geophysical inversion, in general and gravity inversion in particular faces mainly two problems, viz., heavy dependance on quality of initial guess and stability. The proposed method developed on concepts of tunnelling (Levy et al, 1982) and Tikhonov's regularisation tackles both these problems in a overdetermined case of gravity inversion. The algorithm is a recursive one, which progresses through a sequence of local minima of systematically decreasing values. The results of simulation in gravity case do indicate that even if the initial guess is far away from the true solution, the terminal solution reached, is indeed close to true one. Thus, efficacy of joint application of reglarisation and tunnelling is illustrated.

INTRODUCTION

Inverse problems are difficult to solve for many reasons, Viz., Formulation, parametrization, non-uniqueness, instability and heavy dependance on a priori solution. The intent of this paper is to tackle the last problem.

Barring few noteworthy exceptions (Kirkpatrick et al, 1983; Rothman 1985) many of the proposed inversion algorithms suffer from ' Local Minimum Syndrom' i. e., Heavy dependance on initial guess. The only global optimi zation technique actually used to solve a geophysical problem Viz., that of residual statics (Rothman, 1985) is statistical in nature. However, the scheme used leaves much to be desired like separate assessment of a priori distributions for individual problems. Better statistcal schemes for global optimization exist (Price, 1978). However, we prefer a deterministic scheme, which is described in the next section.

THE TUNNELLING ALGORITHM

The tunnelling algorithm (Levy and Montalvo, 1977, 1980; Levy et al, 1982; Levy and

Gomez, 1980, 1985) goes through a series of local minima, each no worse and actually better than the previous one. The chief advantage is that the method doesnot encounter all the local minima. The idea is to start at a

point x_1^0 and from there reach the local minimum x_1^* , then by "tunnelling" (to be explained presently) reach another point x_2^0 such that $f(x_2^0) \leqslant f(x_1^0)$ where $f(x)$ is the objective function, and from there another local minimum x_2^* , and so on. Thus, we have a sequence of points x_1^0 , x_1^* , x_2^0 , x_2^* , ... x_i^0 , x_i^* , ... ,wherein x_i^* is a local minimum and $f(x_i^*) \leqslant f(x_{i-1}^*)$, so that eventually a global minimum is reached. This is illustrated in Fig. 1.

The algorithm employs a tunnelling function,

$$T_1(x, F_i) = \frac{f(x) - f(x_i^*)}{[(x-x_i^*)^T(x-x_i^*)]^\lambda} \tag{1}$$

where F_i is tunnelling parameter vector,

$$F_i = (x_i^*, \lambda, f(x_i^*)) \tag{2}$$

and superscript T is transposition.

Now, the zero of $T_1(x, F_i)$ is sought to find a nominal point x_{i+1}^* , at which next minimization cycle will begin. In such an effort, the role of numerator of Eq.(1) is to seek an argument x (Initially set

as $x = x_i^* + \varepsilon$, ε being a small random vector with $11\varepsilon11 << 1$) with same function value as at last minimum x_i^* and that of denominator is to exclude x_i^* from search by creation of pole at $x = x_i^*$ with pole strength λ (Fig. 2) .

THE REGULARISING TUNNELLING ALGORITHM

Utilising the latitude provided by the tunnelling algorithm as proposed by Levy et al (1982), we have used a variation of our own, which has not been previously proposed or tested. The details are as follows:

A class of gravity inverse problems can be written in terms of a set of overdetermined non- linear equations,

$$h_i(x) = b_i \tag{3}$$

where b_i are Bouguer anomalies at observation points indexed by i, expressed as non-linear functions h_i of the parameter vector x of the geophysical model of the anomaly. In operator form, Eqn. 3 can be written as

$$Hx = 0 \tag{4}$$

cannot be satisfied, one wants to minimize $f(x) = 11\,H\,x\,11^2$. But this is a multi-extremal problem requiring global optimization.

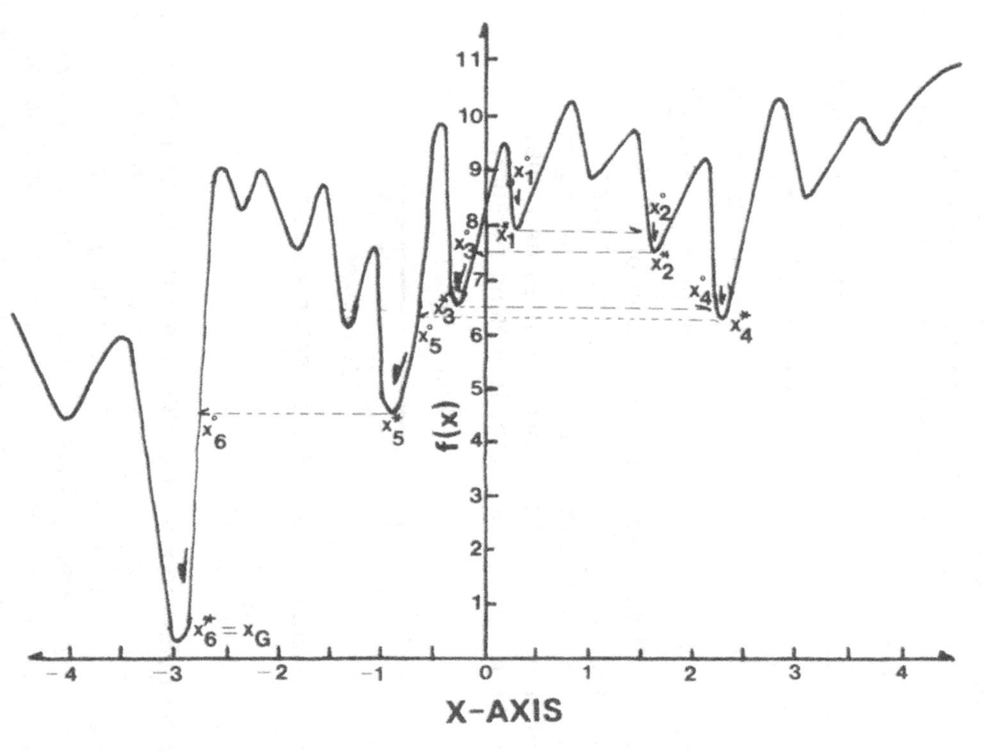

Fig. 1 Illustration of tunnelling in global minimization problem.

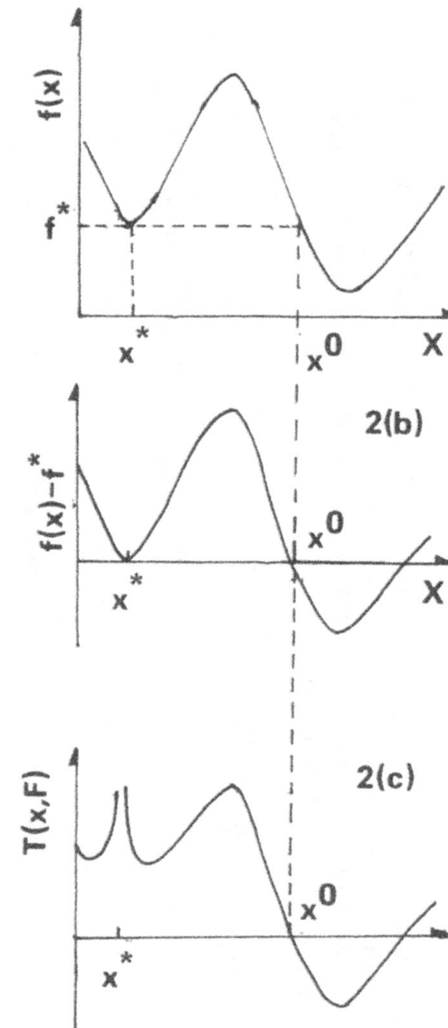

Fig. 2 Concept of tunnelling
   a) Finding a local minimum, x_i^* (shown as x^*)
   b) Seeking x^0, zero of $f(x)-f(x_i^*)$
   c) Creation of a pole of pole strength λ, at x_i^*

For the local minimization, Tikhonov Regularization is used. i.e., Objective function

$$M_\alpha < \underline{x}, \underline{x}^0, H> = 11\, H\,\underline{x}11^2 + \alpha\, 11\, \underline{x} - \underline{x}^0\, 11^2 \qquad (5)$$

is minimized, where 11.11 is the Eucledian norm, $\underline{x}^0$, a priori solution and α is a regularization parameter.

The minimization of M in Eq. (5) leads to the following (Starostenko and Zavaretko, 1976) iterative scheme:

$$[\,\underline{H}'^*(\underline{x}_i^n)\,\underline{H}'(\underline{x}_i^n) + \alpha_p\,\underline{I}\,]\,\underline{x}_i^{n+1} =$$

$$-\underline{H}'^*(\underline{x}_i^n)\,H(\underline{x}_i^n) + \underline{H}'^*(\underline{x}_i^n)\,\underline{H}'(\underline{x}_i^n) + \alpha_p\,\underline{x}^0$$

$$n = 0, 1, 2, \ldots$$

$$\alpha_p = \mu\,\alpha_{p-1}, \mu < 1, p = 0, 1, 2, \ldots \qquad (6)$$

where the superscript n denotes the iteration index within the local minimization phase, subscript i denotes the index in the alternating sequence of local minimization and tunnelling phases and p is an index according to which the regularization parameter is progressively reduced, separate notation for this index providing the facility that this reduction need not be sychronous with the iterative search for the local minimum. Further,

$\underline{H}'(\underline{x}_i^n)$ and $\underline{H}'^*(\underline{x}_i^n)$ are gradient and conjugate gradient operators respectively at the n th iteration, $\underline{I}$ is the identity matrix. When $\underline{x}_i^{n+1} = \underline{x}_i^n$, the local minimization phase is truncated and the resultant x_i^{n+1} is taken to be x_i^*, the starting point for the tunnelling phase. The approximate equality of $\underline{x}_i^{n+1}$ and $\underline{x}_i^n$ is judged in terms of a threshold on the Chebyshev or uniform norm of $\underline{x}_i^{n+1} - \underline{x}_i^n$.

Regularization of the local minimization phase adopted in this paper is an innovation, which was not forbidden by the tunnelling algorithm , because it doesnot prescribe how the local minimization is to be achieved.

As per Eq. (1), here we have a vector tunnelling function

$$T^{(i)}(\underline{x}, \underline{F}_i) = \frac{H(\underline{x}) - H(\underline{x}_i^*)}{[\,(\underline{x}-\underline{x}_i^*)^T(\underline{x}-\underline{x}_i^*)\,]^\lambda} \qquad (7)$$

The task of tunnelling phase is to solve a vector equation

$$T^{(i)}(\underline{x}, \underline{F}_i) = \underline{0} \qquad (8)$$

This is again an overdetermined problem, which can be tackled by minimizing $11\, T^{(i)}\, 11^2$. Thus, the two phases solve the same problem but for different objective functions. This notion of tunnelling vector, thus reduces the programming effort and makes Tikhonov Regularization applicable to tunnelling phase also, thereby enhancing the stability of the entire algorithm.

To check the validity , numerical experiments have been devised, the details of which are discussed in the next sections.

Table 1 Convergence behaviour of Regularising Tunnelling Algorithm

S.No.	Gravity models (Unknowns are Prisms' lateral coordinates)	Convergence		Remarks
		Ordinary method	Our method	
1.	Four Parameters 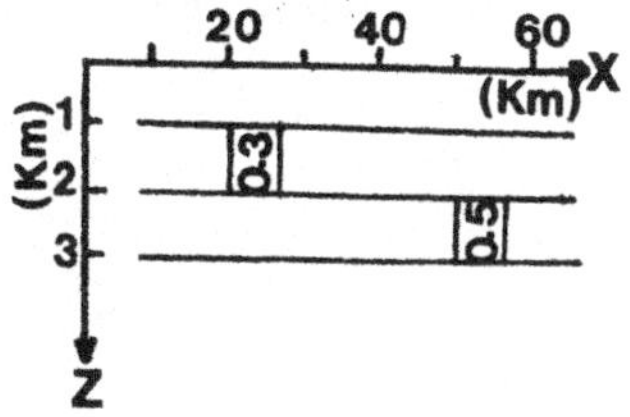 Start Guess: (5, 8, 15, 20) True Solution: (20, 25, 50, 54) l_1-Norm deviation: 102	Not reached even in 250 iterations	50 iterations	Objective function plots (Fig. 4 & 5) are considered to demonstrate the effectiveness of our algorithm.
2.	Seven Parameters 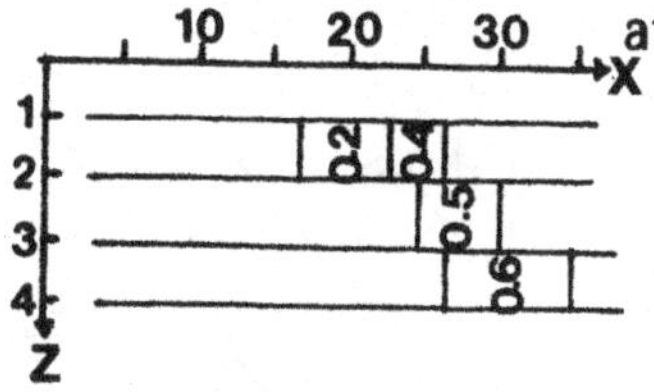 Start Guess: (13, 25, 29, 29, 26, 30, 39) True Solution: (17, 23, 26, 25, 30, 26, 35) l_1-Norm deviation: 25	Not attained	5 Iterations.	
3.	Eleven Parameters 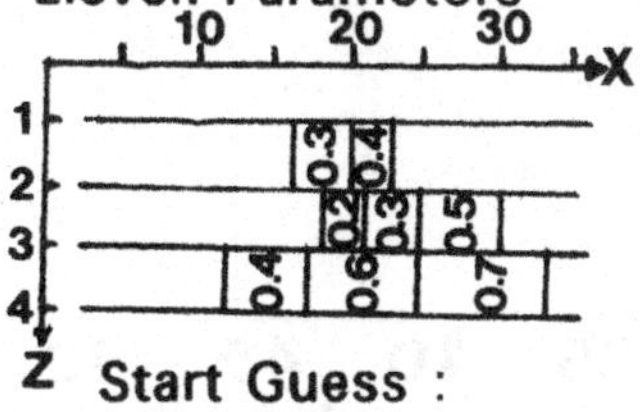 Start Guess: (14, 17, 21, 20, 23, 23, 34, 11, 19, 22, 33) True Solution: (16, 19, 22, 18, 21, 25, 30, 12, 17, 25, 33) l_1-Norm Deviation: 21	Not attained	5 Iterations	Adjustment of tunnelling parameters needed between cycles.

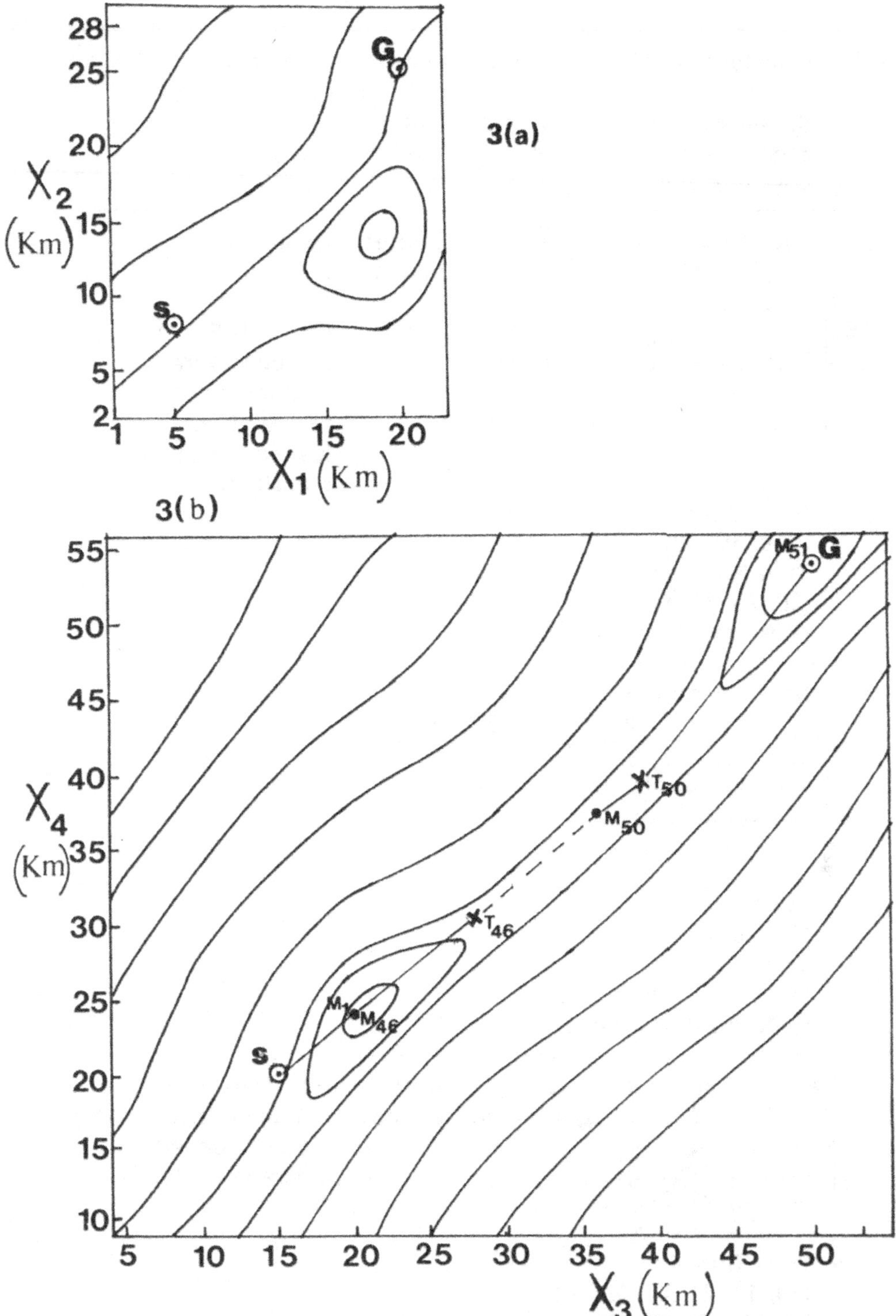

Fig. 3 Misfit function (least–squares error) plots for a 4–parameter model.
a) Projection on X_1X_2 plane (Starting guess stage).
b) Projection on X_3X_4 plane (Starting guess stage).
(The overlay in Fig. 3(b) indicates algorithm's progress from
start guess, S to the global minimum, G through local minima
M_i and tunnel points T_i, i=1, 2, ... in X_3X_4 plane)

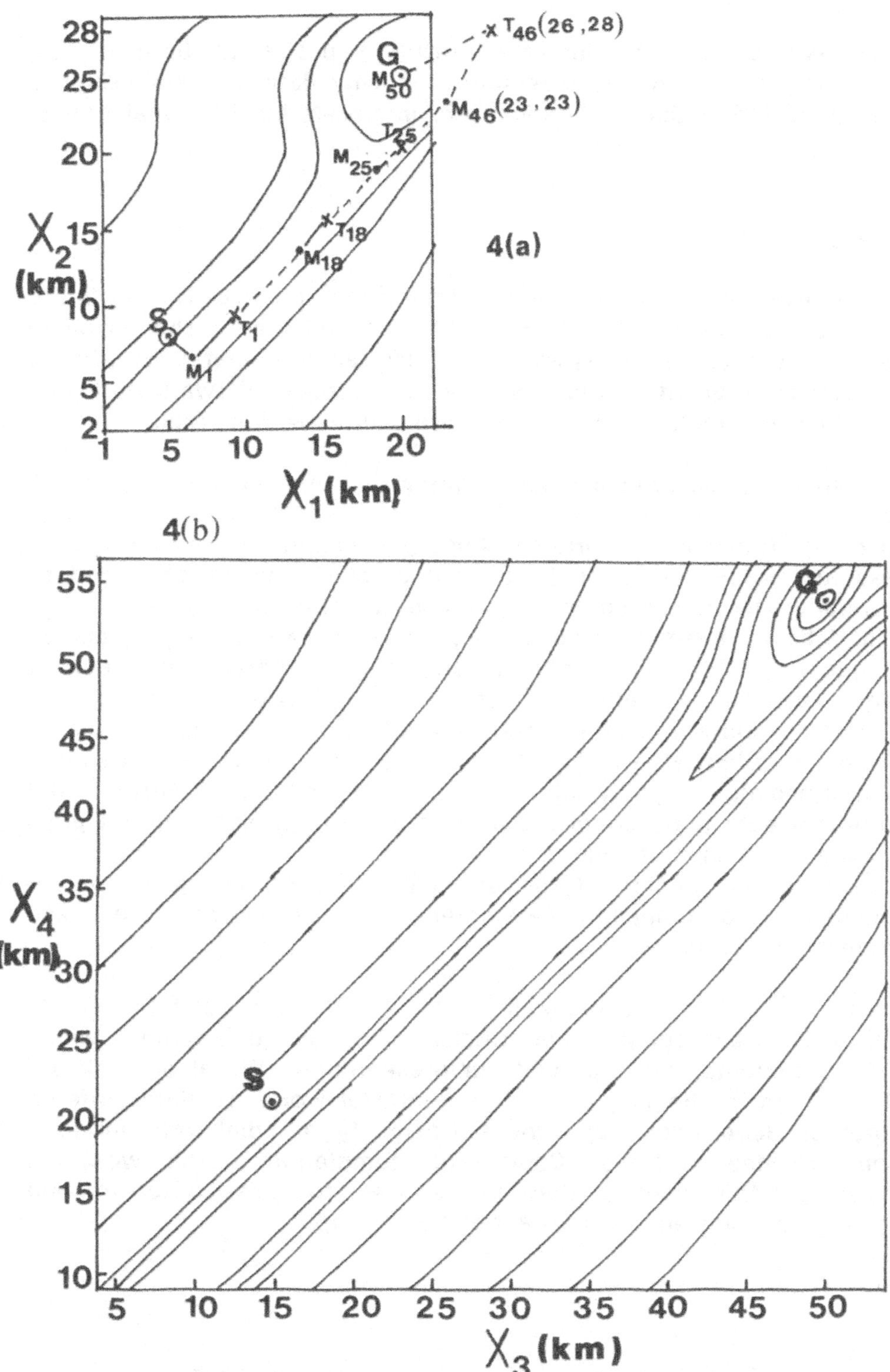

Fig. 4 Misfit function (least-squares error) plots for a 4-parametre model.
   a) Projection on X_1X_2 plane (Terminal solution stage).
   b) Projection on X_3X_4 plane (Terminal solution stage).
      (The overlay in Fig. 4(a) indicates algorithm's progress from
      start guess, S to the global minimum, G through local minima
      M_i and tunnell points T_i, i= 1, 2, ... in X_1X_2 plane.

NUMERICAL EXPERIMENTS AND RESULTS

The purpose of present section is to illustrate with the help of synthetic examples, the application of our algorithm to the inversion of gravity data for establishing a mathematical technique rather than geological inr5 interpretation of actual gravity data.

Models for convergence tests :

To tackle overall convergence, several two- dimensional prisms with different density contrasts located in parallel beds have been chosen. The involved nonlinear inverse problem concerns estimation of lateral coordinates of prism's boundaries from respective gravity profiles across the sources. Forward problems are carried out by Talwani's method. The results are summarized in Table 1 .

The following points emerge after a close scrutiny of achieved results (Table 1 and Fig. 3 & Fig. 4):

a) The Regularising Tunnelling algorithm converges inspite of a poor quality initial guess for multi- dimensional extremal problems, when compared to ordinary inversion algorithms, employing simple minimization schemes.

b) Fig. 3 & Fig. 4 being misfit function projections on $X_1 X_2$ and $X_3 X_4$ planes for initial guess and terminal solution stages(four parameter problem) respectively, illustrate the effectiveness of our algorithm. As can be observed, local minima are absent in terminal solution stage(Fig. 4) , where as their existence is indicated in Fig. 3 (Start guess stage), as well as similar plots for intermediate stages,which are not presented here. Further, the overlay algorithm path plots confirm that local minima presence donot pose a threat for a global minimum search.

c) With a raise in dimensionality of the problem and for poor quality initial guesses, adjustment of tunnelling parameters is necessary, as is the case with 11-dimensional problem.

Acknowledgements: The authors are thankful to Dr. A. Tarantola and Dr. D. N. Rothman for their incisive and constructive criticism of our earlier attempts, which has sharpened our arguments and improved our presentation. We also thank our student, Mr. Mohan Ram for helping us in computational work. Facilities offered to the first author at Geophysics Department, University of Edinburgh during a visiting fellowship provided by British Council for completion of this work are gratefully acknowledged.Our special thanks are due to Dr.V.R.S.Hutton and Dr.R.G.Hipkin for their constant encouragement and kind help.

REFERENCES

Kirkpatrick, S., Galatt, Jr. C. D. and Vecchi, M. P. (1983). Optimization by simulated annealing, *Science,* 220,671–680.

Levy, A. V., Montalvo, A., Gomez, S. and Calderon, A. (1982). Topics in Global Optimization, *Lecture Notes in Mathematics,* Hennart, J. P (ed.),No. 909, 18– 33, Springer– Verlag, Berlin.

Levy, A. V and Gomez, S. (1980) The tunnelling gradient- restoration algorithm for the global minimization of non- linear functions subject to non- linear inequality constraints, IIMAS- UNAM, *Communicaciones*

Tecnicas, Serie Naranja: Investigaciones, No. 231.

Levy, A. V and Gomez, S. (1985) The tunnelling method applied to global opti
mization, *Numerical Optimization 1984,* Boggs,P.T.,Byrd,R. H.
and Schnabel, R. B. (eds.), SIAM, Philadelphia, 213- 214.

Levy, A. V and Montalvo, A. (1977) The tunnelling algorithm for the global
optimization of functions, *Dundee Bienn. Conf. Num. Ana,* Dundee.

Levy, A. V. and Montalvo, A. (1980) A modification to the tunnelling
algorithm for finding the global minima of an arbitrary
one- dimensional function , *Communicaciones Tecnicas,* Seria
Naranja, No. 240, IIMAS- UNAM.

Price, W. L. (1978) A controlled random research procedure for global opti
mization, *Towards Global Optimization 2,* Dixon, L. C. W. and
Szego, G. P (eds.) ,North Holland, Amsterdam, 71- 84.

Rothman, R. H. (1985) Nonlinear inversion, statistical mechanics and
residual estimation, *Geophy.,* 50, 2784- 2796.

Starostenko, V. I. and Zavaretko, A. N. (1976) Application of regularising
algorithm to non-linear inverse problem of gravity - Methodology and
results, *Geophizicheski Sbornik,* 71, 29- 40.

Symposium 103 Author Index

Made in the USA
Monee, IL
07 July 2026